ALSO BY MICHAEL SHERMER

Skeptic: Viewing the World with a Rational Eye

*The Moral Arc: How Science and Reason Lead Humanity
Toward Truth, Justice, and Freedom*

*The Believing Brain: From Ghosts and Gods to Politics and
Conspiracies—How We Construct Beliefs and Reinforce Them as Truths*

*The Mind of the Market: How Biology and Psychology Shape
Our Economic Lives*

Why Darwin Matters: The Case Against Intelligent Design

Science Friction: Where the Known Meets the Unknown

*The Science of Good and Evil: Why People Cheat, Gossip, Care,
Share, and Follow the Golden Rule*

The Borderlands of Science: Where Sense Meets Nonsense

*Denying History: Who Says the Holocaust Never
Happened and Why Do They Say It?* (with Alex Grobman)

In Darwin's Shadow: The Life and Science of Alfred Russel Wallace

How We Believe: Science, Skepticism, and the Search for God

*Why People Believe Weird Things: Pseudoscience,
Superstition, and Other Confusions of Our Time*

The Skeptic Encyclopedia of Pseudoscience (General Editor)

Secrets of Mental Math (with Arthur Benjamin)

HEAVENS ON EARTH

HEAVENS

on

EARTH

THE SCIENTIFIC SEARCH FOR
THE AFTERLIFE,
IMMORTALITY, AND UTOPIA

MICHAEL SHERMER

HENRY HOLT AND COMPANY   NEW YORK

Henry Holt and Company
Publishers since 1866
175 Fifth Avenue
New York, New York 10010
www.henryholt.com

Henry Holt® and 🏛® are registered trademarks of
Macmillan Publishing Group, LLC.

Library of Congress Cataloging-in-Publication Data

Names: Shermer, Michael, author.
Title: Heavens on earth : the scientific search for the afterlife, immortality, and
    utopia / Michael Shermer.
Description: First edition. | New York : Henry Holt and Co., 2018. | Includes index.
Identifiers: LCCN 2017033312 (print) | LCCN 2017043333 (ebook) | ISBN
    9781627798563 (eBook) | ISBN 9781627798570 (hardback)
Subjects: LCSH: Immortality. | Future life. | Immortalism. | Religion and science.
Classification: LCC BL530 (ebook) | LCC BL530 .S53 2018 (print) | DDC
    202/.3—dc23
LC record available at https://lccn.loc.gov/2017033312

Our books may be purchased in bulk for promotional, educational, or business use. Please
contact your local bookseller or the Macmillan Corporate and Premium Sales Department at
(800) 221-7945, extension 5442, or by e-mail at MacmillanSpecialMarkets@macmillan.com.

First Edition 2018

Designed by Kelly S. Too

Printed in the United States of America

1 3 5 7 9 10 8 6 4 2

To

Vincent Richard Walter Shermer

That you may find your own heaven on earth . . . by looking within

Therefore death, the most terrifying of evils, is nothing to us, since for the time when we are, death is not present; and for the time when death is present, we are not. Therefore it is nothing either to the living or the dead since it is not present for the former, and the latter are no longer.

—Epicurus, *Letter to Menoeceus*, third century B.C.E.[1]

CONTENTS

PART IV: MORTALITY AND MEANING

HEAVENS ON EARTH

MEMENTO MORI

Life is short, and shortly it will end;
Death comes quickly and respects no one,
Death destroys everything and takes pity on no one.
To death we are hastening, let us refrain from sinning.

—*Ad mortem festinamus* ("We hasten toward death"),
medieval memento mori, *Libre Vermeil de Montserrat*, 1399

Between 50,000 B.C.E. and 2017 C.E. about 108 billion people were born.[2] There are alive today around 7.5 billion people. This makes the ratio of the dead to the living 14.4 to 1,[3] which means that only 7 percent of everyone who ever lived is alive today.[4] Of those 100.5 billion people who have come and gone, not one of them has returned to confirm the existence of an afterlife, at least not to the high evidentiary standards of science.[5] This is the reality of the human condition. Memento mori— "Remember that you have to die."

Life is short. Thanks to public health measures and medical technologies, life expectancy has more than doubled and is now approaching age 80 for Westerners, but no one has exceeded the maximum lifespan of approximately 125 years for our species.[6] The current record of 122 years, 164 days is held by Jeanne Calment of France (1875–1997), although some poorly documented claims are made for longer-lived people, so I set the upper ceiling at 125. While I was writing this book the world's oldest person died at the age of 116, replaced by another centenarian, also aged 116.[7] This cycling through of the longest-lived person will continue indefinitely, but unless there are major medical and technological breakthroughs in life extension, which we will consider in due course, it is very unlikely to exceed 125. Memento mori.

Life is final. The poet Dylan Thomas urged us "Do not go gentle into that good night," but instead "Rage, rage against the dying of the light." Most people, though, opt for John Donne's conviction that "One short sleep past, we wake eternally."[8] But to get there you have to die. Memento mori.

The belief that death is not final is overwhelmingly common. Since the late 1990s, the Gallup polling group has consistently found that between 72 and 83 percent of Americans believe in heaven.[9] A 1999 study found that Protestants remained steadfast in their heavenly belief at 85 percent over the decades, whereas afterlife belief among Catholics and Jews increased from the 1970s to the 1990s.[10] A 2007 Pew Forum survey found that 74 percent of all Americans believe heaven exists, with Mormons topping the chart at 95 percent.[11] A 2009 Harris poll found that 75 percent of Americans believe in heaven, ranging from a low of 48 percent for Jews to a high of 97 percent for born-again Christians.[12] Tellingly, belief in the devil and the invocation of hell has been in gradual decline in both liberal and conservative churches,[13] and in all polls belief in hell trails belief in heaven by 20 to 25 percent, thereby confirming the overoptimism bias.[14] Globally, rates of belief in heaven in other countries typically lag behind those in America, but they are nonetheless robust. A 2011 Ipsos/Reuters poll, for example, found that of 18,829 people surveyed across 23 countries, 51 percent said they were convinced that an afterlife exists, ranging from a high of 62 percent of Indonesians and 52 percent of South Africans and Turks down to 28 percent of Brazilians and only 3 percent of the very secular Swedes.[15]

So powerful and pervasive are such convictions that even a third of agnostics and atheists proclaim belief in an afterlife. Say what? A 2014 survey conducted by the Austin Institute for the Study of Family and Culture on 15,738 Americans between the ages of 18 and 60 found that 13.2 percent identify as atheist or agnostic, and 32 percent of those answered in the affirmative the question "Do you think there is life, or some sort of conscious existence, after death?"[16] The percentage is certainly lower than the overall mean of 72 percent for all Americans in this study, but it is surprisingly high given our understanding of the worldview held by most atheists and agnostics, which commonly presumes that if there is no God then there is no afterlife. Perhaps that is presumptuous; who knows what is in the minds of people when they complete such surveys? But given the fact that 6 percent of atheists and agnostics also believe in the bodily resurrection of the dead (compared to 37 percent overall), perhaps belief in God and immortality are orthogonal—independent of each other. One may believe in an afterlife but not God. Or both. Or neither.

DYING TO GO TO HEAVEN

Heavens above may or may not be real, but heavens on earth are, at least in the minds of those who believe in them. In that sense, the empyrean realm of gods and heavens that resides in the brains of believers is as real as anything in the terrestrial kingdom. Given the power of beliefs to drive people to act, we should treat such attitudes as seriously as we would political, economic, or ideological beliefs, which hold the same power over actions. As the Saudi cleric Abdullah Muhaisini shouted to his rebel factions in Syria to exhort them to retake the besieged city of Aleppo in 2016, referring to the paradise filled with beautiful women with lustrous eyes with which they would be rewarded upon death:

> Where are those who want 72 gorgeous wives? A wife for you, O martyr in heaven, if she spits in the sea, the sea becomes sweet. If she kisses your mouth, she fills it with honey . . . If she sweats, she fills paradise with perfume. Then how would it be in her embrace?[17]

Ever since 9/11, people in the West have become understandably curious about the role of heavenly beliefs in suicide terrorist attacks. Although most Muslim scholars say that the Qur'an forbids suicide—much less suicidal bombings that kill civilians—there are obviously work-arounds for this proscription, given the proliferation of young men (and a few women) intent on becoming martyrs by donning bomb vests and blowing themselves up in crowded public places. In fact, in Islam the only people allowed to skip the purgatory-like judgment stage and go directly to paradise are martyrs. According to the religious scholar Alan Segal, "in a 'holy war,' the *mujahidin* can attain the status of the *shahid*, the martyr. Not only that, the early *Hadith* literature encourages martyrdom. The person seeking martyrdom, the *talab al-shahada*, is to be exalted and emulated. This kind of martyrdom is earnestly prayed for and devoutly wished for."[18]

It was in fact Muhammad himself who ruled that as a general principle any Muslim soldier who died while attacking an infidel would go straight to paradise. Of course he would say that, given how well the promise motivated his own troops on March 15, 624, when Muhammad's army faced a vastly larger force at the battle of Badr. After a lengthy prayer vigil, Muhammad announced to his anxious soldiers that the archangel Gabriel told him that an entire angelic force would be on their side and that anyone killed that day would instantly wake up in paradise. According to legend, a fifteen-year-old soldier named Umayr proclaimed in response: "Wonder of wonders! Is there nothing between me and my entry into paradise but that these men kill me?" Muhammad's force won the battle and allegedly suffered only fourteen casualties that day, one of whom was, ironically (or not), Umayr. As in the Wild West, when the legend becomes fact, print the legend.[19]

To carry over this martial payoff into modern suicidal missions, the "enemy soldiers" to be defeated are the invading armies of the Great Satan—Israel and America—against which those who self-identify as Muslim martyrs are fighting. For this small but noticeable minority of the world's Muslims, by definition, anyone who supports Israel or the United States is an infidel, and in this context, any violent act committed against the Great Satan is done in self-defense. The Satanic West, then, is anti-Islamic by definition. Thus, this form of terrorism differs from those of the political anarchists of the early twentieth century and the Marxist revolutionaries of the late twentieth century in that Islamic terrorists are willing to die not just for a political cause, but for religious motives with the promise of paradise as the reward. A paradigmatic statement of this modern belief comes from the 9/11 hijacker Mohammed Atta, whose suicide note (found in the luggage that he left in his rental car that morning before flying American Airlines flight 11 into the World Trade Center building) included the following passage:

> When the confrontation begins, strike like champions who do not want
> to go back to this world. Shout, "Allahu Akbar," because this strikes
> fear in the hearts of the nonbelievers. Know that the gardens of paradise

are waiting for you in all their beauty, and the women of paradise are
waiting, calling out, "Come hither, friend of God." They have dressed
in their most beautiful clothing.[20]

This religious conviction was reinforced in a 2016 article in the ISIS
publication *Dabiq* titled "Why We Hate You, Why We Fight You," in
which six reasons are enumerated:[21]

1. We hate you, first and foremost, because you are disbelievers;
   you reject the oneness of Allah—whether you realize it or not—
   by making partners for Him in worship, you blaspheme against
   Him, claiming that He has a son.
2. We hate you because your secular, liberal societies permit the
   very things that Allah has prohibited while banning many of the
   things He has permitted.
3. In the case of the atheist fringe, we hate you and wage war
   against you because you disbelieve in the existence of your Lord
   and Creator.
4. We hate you for your crimes against Islam and wage war against
   you to punish you for your transgressions against our religion.
5. We hate you for your crimes against the Muslims; your drones and
   fighter jets bomb, kill, and maim our people around the world.
6. We hate you for invading our lands and fight you to repel you and
   drive you out.

The unnamed author reminds his readers not to be thrown off by
the secondary political motives. "The fact is, even if you were to stop
bombing us, imprisoning us, torturing us, vilifying us, and usurping our
lands, we would continue to hate you because our primary reason for
hating you will not cease to exist until you embrace Islam." And: "What's
equally if not more important to understand is that we fight you, not
simply to punish and deter you, but to bring you true freedom in this life
and salvation in the Hereafter."

There it is. The Hereafter. Whatever other motives these suicide
terrorists may have for committing violence—money, sex, adventure,
U.S. foreign policy[22]—anyone who doubts the sincerity of their deep
religious conviction that they will be rewarded in heaven for their
murderous martyrdom is living in denial.

HEAVENS' PLURALITY

The many and varied beliefs about the afterlife and immortality are the reason this book's title—*Heavens on Earth*—is pluralized, and the earthly genesis of such beliefs is indicative of their origin in human nature and culture. This book is about one of the most profound questions of the human condition, one that has driven theologians, philosophers, scientists, and all thinking people to try to understand the meaning and purpose of our life as mortal beings and discover how we can transcend our mortality. It is about how the awareness of our mortality and failings has led to beliefs in heaven and hell, in afterlives and resurrections both spiritual and physical, in utopias and dystopias, in progress and decline, and in the perfectibility and fallibility of human nature. There are nearly as many ideas about heaven—and heavens on earth— as there are people who have thought seriously about the matter of what happens after we die and what we can do to perfect life while we're alive. This transcendence leads to a quest for spiritual immortality in heaven, physical immortality on earth, and the perfectibility of society here and now.

The luminaries we will meet in this book include psychologists and anthropologists and their theories about death and dying and how the awareness of our mortality affects us; archaeologists and historians on who were the first people to become aware of their own mortality and how this awareness led to the creation of myths and religions; Jews, Christians, and Muslims and their monotheistic ideas about heaven and hell, the resurrection of both body and soul, and what happens after we die; spiritual seekers from other religious traditions who seek immortality through altered states of consciousness, including modern spiritual gurus like Deepak Chopra and their belief in transcendent consciousness for eternal life; cognitive scientists in search of explanations for anomalous psychological experiences, and psychic mediums who believe that we can talk to the dead; scholars and scientists who treat near death experiences and the belief in reincarnation as evidence of the afterlife, and skeptics who interpret them from a more materialistic perspective; secular philosophers and scientists in search of immortality through radical life extension, minimal senescence, antiaging remedies, cryonics, transhumanism lifestyles, singularity technologies, computer mind uploading, and other afterlives for atheists; imaginative writers who envision perfect societies; dreamers who attempt to construct utopias; pessimists who lament the decline of civilization; dictators and demagogues who exploit these fears and attempt to rebuild societies in their own imagined fashion of what a paradisiacal state should be, only to see it collapse after the inevitable collision with reality—thus do utopian dreams turn into dystopian nightmares.

Finally, at the end of this voyage we will consider such ultimate problems as why are we mortal and how our species can become immortal, what it means if there is no heaven above or here on earth, and how we can find meaning in an apparently meaningless universe. There are scientific answers to such deep questions, if we reflect upon them with reason, honesty, and courage.

PART I

VARIETIES OF
MORTAL EXPERIENCES AND
IMMORTAL QUESTS

To be immortal is commonplace; except for man, all creatures are immortal, for they are ignorant of death; what is divine, terrible, incomprehensible, is to know that one is mortal.

—Jorge Luis Borges, *The Immortal*, 1943

CHAPTER 1

A LOFTY THOUGHT

Imagining Mortality

> Never to have been born at all:
> None can conceive a loftier thought!
> And second-best is this: Once born,
> Quickly to return to the dust.
>
> —Sophocles, *Oedipus at Colonus*, 406 B.C.E.[1]

Where were you before you were born?

Come again? This question strikes most of us as nonsensical, because we didn't exist before we were born. The same problem arises in imagining your death. Try it. What comes to mind? Do you see your body as part of a scene, perchance presented in a casket surrounded by family and friends at your funeral? Or maybe you see yourself in a hospital bed after expiring from an illness, or on the floor of your home following a fatal heart attack? None of these scenarios—or any others your imagination might conjure—are possible, because in all cases, in order to observe or imagine a scene you must be alive and conscious. If you are dead you are neither. You can no more visualize yourself after you die than you can picture yourself before you were born.

Existence doesn't just precede essence, as Jean-Paul Sartre conjectured in one of the founding documents of the existentialist movement.[2] Existence *is* essence. No existence, no essence. As the German poet and philosopher Johann Wolfgang von Goethe framed the problem, "It is quite impossible for a thinking being to imagine nonbeing, a cessation of thought and life. In this sense everyone carries the proof of his own immortality within himself."[3] Sigmund Freud reflected on death in a similar vein: "We cannot, indeed, imagine our own death; whenever we try to do so we find that we survive ourselves as spectators."[4]

To experience something, you must be alive, so we cannot personally experience death. Yet we know it is real because every one of the hundred billion people who lived before us is gone. That presents us with something of a paradox.

THE MORTALITY PARADOX

In his now classic Pulitzer Prize–winning 1973 book *The Denial of Death*, the anthropologist Ernest Becker oriented our dualistic place in nature,

> up in the stars and yet housed in a heart-pumping, breath-grasping body that once belonged to a fish and still carries the gill-marks to prove it. Man is literally split in two: he has an awareness of his own splendid uniqueness in that he sticks out of nature with a towering majesty, and yet he goes back into the ground a few feet in order blindly and dumbly to rot and disappear forever. It is a terrifying dilemma to be in and to have to live with.[5]

Is it terrifying? I don't think it is, but many people do. In his book *Immortality*, for example, the British philosopher Stephen Cave contends that the attempt to resolve the paradox of being aware of our own mortality and yet not being able to imagine nonexistence has led to four immortality narratives: (1) *Staying Alive*: "like all living systems, we strive to avoid death. The dream of doing so forever—physically, in this world—is the most basic of immortality narratives." (2) *Resurrection*: "the belief that, although we must physically die, nonetheless we can physically rise again with the bodies we knew in life." (3) *Soul*: The "dream of surviving as some kind of spiritual entity." (4) *Legacy*: "More indirect ways of extending ourselves into the future" such as glory, reputation, historical impact, or children.[6] Cave's four-part schema is instructive, so a brief overview is in order to resolve the paradox provisionally.

First, *staying alive* is not presently possible. There are scientists working to extend our upper age ceiling through various medical technologies, but for now the bookmakers' odds-on bet is that no one alive today will live beyond 125 years. Even if medical science raises the age roof by a few years or decades, the dream of living centuries or millennia is a vaporous one.

Second, *resurrection* harbors two logical problems with both religious and scientific forms of reconstituting your body: (1) *The Transformation Problem*: How could you be reassembled just as you were and yet this time be invulnerable to disease and death? To avoid these problems you would need to be resurrected in a much different state than you are now, so this new identity would not really be you. One work-around is to preserve your *connectome*—the brain's equivalent to the genome—where your thoughts, memories, and "self" are stored, and then perhaps upload all that information into a computer. I am involved in one aspect of this research and will discuss it at length in chapter 7, but for here I will note that in addition to the technological hurdles, this option leads to a second difficulty. (2) *The Duplication Problem*: How would duplicates be different from twins? That is, even if a godlike supercomputer in the far future had virtually limitless digital power to make a perfect copy of you, it would be just that—a copy with the same thoughts and memories as you until it began its own independent existence. At that point your copy will have separate life experiences and memories, and you and your copy would thus be logically indistinguishable from identical twins, whom we legally treat as autonomous persons and not as duplicates of the same individual.

Third, the *soul* has been traditionally conceived as a separate entity ("soul stuff") from the body, but neuroscience has demonstrated that the mind—consciousness, memory, and the sense of self representing "you"—cannot exist without a brain. When portions of the brain die as a result of injury, stroke, or Alzheimer's, the corresponding functions we call "mind" die with them. No brain, no mind; no body, no soul. Scientists working to preserve the connectome are also considering techniques to either reawaken a frozen brain with its connectome intact (cryonics), or scan every last synapse in a brain and digitize it so that it can be "read" like a book or reawakened in a computer. This *scientific soul* would be the first form of soul stuff ever measured, but as we shall see, the obstacles to achieving this form of immortality are beyond extraordinary. I don't think this will happen in my lifetime, or perhaps anyone's lifetime, leaving us with . . .

Fourth, *legacy* isn't strictly a form of immortality at all, but more of a type of memory—the remembrance of a life—and as Woody Allen quipped: "I don't want to be immortal through my work; I want to be immortal by *not dying.* I don't want to live on in the hearts of my countrymen; I want to live on in my apartment."[7] At the moment this is the best we can do, but it's something, given how important our lives can be in the lives of those we know and love (and even those we don't), but it is understandably less emotionally satisfying than our desire to live literally forever.

Cave resolves the paradox by contending that the *legacy* narrative we tell ourselves is the driving force behind art, music, literature, science, culture, architecture, and other artifacts of civilization—and even civilization itself. The legacy driver is *terror*, which has now a full-blown research paradigm called *Terror Management Theory* (TMT), proposed by the psychologists Sheldon Solomon, Jeff Greenberg, and Tom Pyszczynski in numerous scientific papers and more extensively in their book *The Worm at the Core: On the Role of Death in Life.*[8] Inspired by Ernest Becker, the curious title comes from William James's classic 1902 work *The Varieties of Religious Experience*, in which the psychologist conjectured "a little cooling down of animal excitability and instinct, a little loss of animal toughness, a little irritable weakness and descent of the pain-threshold, will bring the worm at the core of all our usual springs of delight into full view, and turn us into melancholy metaphysicians."[9] According to TMT, awareness of one's mortality focuses the mind to produce positive emotions (and creations) to avoid the terror that comes from confronting one's death. Solomon explains the theory:

> Humans "manage" this terror by embracing cultural worldviews—beliefs about reality—shared with other group members to convey to each of us a sense that we are valuable individuals in a meaningful universe, and hence eligible for literal and/or symbolic immortality. Accordingly, people are highly motivated (albeit quite unconsciously) to maintain faith in their cultural worldviews and a confidence in their self-worth (i.e., self-esteem); and threats to cherished beliefs and/or self-esteem instigate defensive efforts to bolster their worldviews and self-esteem.[10]

Thus, we create and invent, build and construct, write and sing, perform and compete, to attenuate the terror of contemplating our own mortality. Civilization is the product not of ambition but of trepidation.

I have my doubts. First, it is not obvious why contemplating death should lead people to experience terror, get defensive about cultural worldviews, or feel the need to bolster self-esteem. It could just as well lead people to feel more sympathy for others who, after all, are in the same existential boat. Second, why wouldn't such despair lead people to just give up on building or creating anything, since it is fruitless in the long run, if not the short? Third, TMT scientists admit that much of their theory depends on *unconscious* states of mind that are notoriously difficult to discern and require subtle priming of the brain to elicit.

TMT proponents even go so far as to conjecture that our Paleolithic ancestors died prematurely from death terror. How? Those hominid groups that developed religious rituals to quell their death terror were more likely to survive. "A creature with the dawning realization of its own mortality and no system of spiritual beliefs to quell the consequent fear would seem unlikely to venture forth and take the risks necessary for their own or their group's survival," Solomon and his colleagues conjecture:

> Hominids with faith in some spiritual protection would be more bold and confident in engaging in the risky tasks necessary for survival in harsh dangerous environments. This suggests that with the dawn of awareness of mortality, hominid groups with particularly compelling spiritual beliefs and individuals particularly capable of sustaining faith in such beliefs would have had adaptive advantages.[11]

It's a colorful story, but one lacking in empirical evidence and not as probable as competing hypotheses of the evolutionary origins of culture and religion and the psychological processes underlying it. Human behavior is multivariate in causality, and fear of death is only one of many drivers of creativity and productivity, if it is one at all. The capacity to reason is a feature of our brain that evolved to form patterns and make connections in the service of survival and reproduction in the environment of our evolutionary ancestry. Reason is part of our cognitive makeup, and once it is in place it can be put to use in analyzing problems it did not originally evolve to consider. The psychologist Steven

Pinker calls this an *open-ended combinatorial reasoning system*, and he notes that "even if it evolved for mundane problems like preparing food and securing alliances, you can't keep it from entertaining propositions that are consequences of other propositions."[12] The capacity to reason and communicate symbolically is employed in hunting, surely a more basic survival skill than the management of death terror. TMT theorists propose that "before venturing out on a hunt or exploring new territory, early *Homo sapiens* may have performed rituals and told stories about how the spirits would help them slay mammoths, leopards, and bears and protect them from potential dangers in the physical world."[13] Maybe, and some interpreters of the prehistoric cave paintings of Altamira, Lascaux, and Chauvet that feature bison, horses, aurochs, and deer attribute these images to hunting magic, but skeptics point out that many of these animals were not hunted in those regions (no bones of these beasts have been found there), and other animals that were commonly hunted (for which there are ample bones in the caves and nearby showing marks of being hunted) are not featured in the cave paintings.[14] In any case, what does symbolic hunting magic have to do with death terror?

A more pragmatic cognitive skill may be at work here, such as that proposed by a professional animal tracker (and historian of science) named Louis Liebenberg, who argues that our ability to reason and communicate symbolically is a by-product of fundamental skills developed by our ancestors for tracking game animals, starting with hypothesis testing. "As new factual information is gathered in the process of tracking, hypotheses may have to be revised or substituted by better ones. A hypothetical reconstruction of the animal's behaviors may enable trackers to anticipate and predict the animal's movements. These predictions provide ongoing testing of hypotheses."[15] The development of tracking also involves the cognitive process called *theory of mind*, or mind reading, in which trackers put themselves into the mind of the animal they are pursuing and imagine what it might be thinking in order to predict its actions.

This, it seems to me, is a far likelier explanation for the evolution of symbolic reason than death terror. Once the neural architecture is in place to deduce, say, that "a lion slept here last night," a person can substitute any other animal or object for "lion" and can swap "here" with "there" and "last night" with "tomorrow night." The objects and time elements of the reasoning process are fungible. As Pinker explains in *How the Mind Works*, this interchangeability is a by-product of neural systems that evolved for basic reasoning abilities such as tracking animals for food.[16] It's a bottom-up combinatorial reasoning process that includes *induction* (reasoning from specific facts to general conclusions) and *deduction* (reasoning from general principles to specific predictions) that allowed humans to scale up from basic survival skills such as hunting and gathering to more abstract concepts such as death, the afterlife, souls, and God. In this sense, then, religion is not a direct adaptation to living conditions but a by-product of these abstract reasoning abilities.

An even more elementary evolutionary driver of creativity and culture is sex and mating—sexual selection, in the parlance of evolutionary theory—in which organisms from bowerbirds to brainy bohemians engage in the production of magnificent works in order to attract mates. Big blue bowerbird nests constructed by males appeal to females, and the bigger and the bluer they are, the more offspring are in the offing. Likewise big-brained bohemians, whose orchestral music, epic poems, stirring novels, monumental architecture, and scientific discoveries may be motivated by the desire to attract mates and gain status. As the evolutionary psychologist David Buss noted in his critique of Terror Management Theory: "TMT is anchored in an outmoded evolutionary biology that stresses survival, but ignores reproduction," it "fails to delineate precisely how the hypothesized psychological mechanisms help humans solve actual adaptive problems of survival and reproduction, and instead focuses nearly exclusively inwardly on psychological protection," it "fails to consider why anxiety itself would have evolved," and it "fails to account for known sex differences in social motivation, death rates, and the causes of death rates."[17] The evolutionary psychologist Geoffrey Miller reinforced the point in his aptly titled book *The Mating Mind*. Those who create and invent, build and construct, write and sing, perform and compete most effectively leave behind more offspring and thus pass their creative genes into future generations.[18] As the belletrist extraordinaire Christopher Hitchens once told me, mastering the pen and the podium means never having to dine or sleep alone.

To this end, I am not at all sure that TMT proponents are even measuring what they think they are measuring in their experiments. In my opinion, the claim that people feel "terror" when contemplating mortality is an assertion, not an observation, and its dependence on unconscious states of mind makes it even more problematic when determining what, exactly, is being tested. "The really tricky thing with theories like this is not what to do with statistical refutations, but rather what to do with supposed statistical confirmations," the psychologist Frank Sulloway told me when I queried him about TMT. "This problem previously arose in connection with psychoanalysis, and Hans Eysenck and others later wrote books showing that those zealous psychoanalytic devotees testing their psychoanalytic claims systematically failed to consider what other theories, besides the one researchers thought they were testing, would also be confirmed by the same evidence." Context is key. "Change the context slightly and one often gets very different results in research on human behavior," Sulloway continued. "So one needs to consider exactly how the context of any statistical test might be altering what you think you are actually testing. This problem is akin to the one about considering what alternative theories are also confirmed by the same evidence."[19]

For example, in a study Sulloway and I conducted on why people say that they believe in God and why they think *other people* believe in God, although we were not testing TMT, we found what I would interpret as results contrary to the centrality of terror in TMT's model.[20] In our survey, in addition to collecting data on personal and family background and religious beliefs and commitments, we asked an open-ended essay question on why respondents believe or disbelieve in God, and why they think *other people* believe or disbelieve in God, which Sulloway and I independently coded, along with an independent judge who was blind to the purposes of the study. Together we assessed all responses as falling into one or more of fourteen belief categories and six nonbelief categories, which we subsequently had coded by a second set of five judges who were also blind to the study's purpose. The twenty categories were then reclassified into one of three summary groups: emotional responses, intellectual responses, and undetermined responses. Figure 1-1 shows the results for the first category, which includes the fear of death.

Note that only 3 percent of our respondents listed "fear of death" or "fear of the unknown" in their reasons for their own belief in God,

Figure 1-1. Emotional Reasons Why People Believe in God and Think Other People Believe in God

GENERAL CATEGORY	COMMON EXAMPLES	SELF	OTHER
Faith	The need to believe in something/faith/just because	13%	16%
Emotion	A sense of God in everyday life	10%	5%
Comfort	Belief is comforting/relieving/consoling	9%	35%
Meaning	Belief gives meaning to life	6%	15%
Fear of Death	Fear of death/fear of the unknown	3%	20%
Morality	Without God there would be no morality	3%	9%
Social Factors	Social aspects, such as peer pressure	1%	17%
Ignorance	Stupidity, lack of education, laziness, moral sloth, responsibility avoidance	0%	15%

which I find revealing in the context of TMT's central tenet; and it is telling that these same people attributed the fear of death or the unknown to *other people's* reasons for belief. TMT may be more revealing of the theorists' projections than their subjects' fears.

The sociologist of religion Kevin McCaffree makes a similar point in putting death fears into context when I queried him about TMT. First, in our evolutionary past, anxiety evolved to direct our attention to survival-relevant concerns, such as hunting, mating, and maintaining a good reputation in one's community. "It is important to notice that these concerns are survival-*relevant*, but they are not concerns about survival (or death) *in itself*. Hunter-gatherers' concerns were more practical." Today, McCaffree continues, "our anxieties are equally practical—car payments, student loans, divorce papers, unemployment and so on. We are certainly motivated to manage these anxieties, but, again, these are anxieties regarding concerns relevant to our survival and flourishing, *not* anxiety concerning survival (or death) in itself." McCaffree also notes that studies show people in highly secularized countries like Sweden and Denmark, where rates of religiosity are among the lowest in the world, seem not to have much death anxiety at all, "not because they love death, but because they understand that there is little they can do about it, and so they choose to focus on aspects of life they can enjoy and exert control over."[21]

WHAT DO PEOPLE THINK ABOUT WHEN FACING DEATH?

While slipping in and out of consciousness as he lay dying of cancer, the Nobel laureate physicist and raconteur Richard Feynman, after a life filled with enough clever sayings and charming stories to fill three volumes,[22] managed only a final utterance of "I'd hate to die twice. It's so boring."[23] Christopher Hitchens came to much the same conclusion about dying as Feynman did, recording his final thoughts while undergoing treatment for esophageal cancer in a series of *Vanity Fair* essays ("Topic of Cancer," "Tumortown"), gathered posthumously in a starkly titled book, *Mortality*. After swiftly dispatching Elisabeth Kübler-Ross's famous (and flawed) stage theory of dying (not everyone goes through all five stages of denial, anger, bargaining, depression, and acceptance, for example, or in that order if they do), Hitch reflected:

> In one way, I suppose, I have been "in denial" for some time, knowingly burning the candle at both ends and finding that it often gives a lovely light. But for precisely that reason, I can't see myself smiting my brow with shock or hear myself whining about how it's all so unfair: I have been taunting the Reaper into taking a free scythe in my direction and have now succumbed to something so predictable and banal that it bores even me.[24]

Sadly, Hitch's end came too soon, and as he told an audience at a public event we both attended shortly before his death, "I'm dying . . . but so are all of you."

Transitioning from life to death through the dying process reminds us of what really matters in life, a point articulated by my first college professor, Richard Hardison, who taught me astronomy, philosophy, and psychology as an undergraduate and who educated me about life

for decades after. He was one of the smartest and most cognitively dex-
terous people I've known, but like so many of his generation (the "silent
generation," born between 1925 and 1945), he was reticent with his
emotions, rarely showing affection even for his closest friends, a personality trait he became painfully cognizant of near the end of his life,
including an observation about it in a farewell letter he penned when
he thought he was dying at age eighty-seven. He recovered and lived
another three years, but at his memorial service another ex-student and
friend, Russell Waters, circulated the letter, which began with Dick's
confession of the awareness of the possibility of his dying in his sleep:
"Strangely, I felt no panic, no dread . . . only a concern that I might be left
without the time to thank friends and family for the many wonderful
things they have done to enhance the quality of my life." His "death aura
passed and the morning dawned as usual," but "this served as a wake-
up call and reminded me that I should write without further delay."
This he did, confessing that it was his friends and family that mattered
most, at the acknowledgment of which "I'm finding that it's already difficult to hold back the tears as I write." His tear-stained letter ends:

> Finally, "love" isn't a word that comes easily to the American male, and
> looking back, my failure to use it a lot more was unfortunate. I should
> have expressed my fondness, and yes, my love, much more often. But at
> least in parting, I can hope that all of you, my friends and family, may
> know of the depth of my appreciation of the prominent role you have
> played in my life.

Love comes more easily to my generation of American males, and so
I am not too taciturn to say that I loved Dick Hardison.

LOVE ON DEATH ROW: TESTING EMOTIONAL
PRIORITY THEORY

Knowing that the end may come sooner rather than later brings death
awareness into sharper focus and motivates the mind to act with clarity
about life's deepest meanings, not out of terror but out of time. And
love. An alternative to Terror Management Theory is one that might be
called *Emotional Priority Theory* (EPT), or the prioritization of one's

emotions when confronted with mortality. As Samuel Johnson noted: "Depend on it, Sir, when a man knows he is to be hanged in a fortnight, it concentrates his mind wonderfully."[25] Facing death focuses one's mind on the most important emotions in life, love being arguably the deepest. In fact, love is so powerful an emotion that it can be addictive, like chocolate and cocaine, and the neurochemistry of it can be tracked.[26] Lust is heightened by dopamine, a neurohormone produced by the hypothalamus and associated with learning and positive reinforcement, which also triggers the release of testosterone, another hormone intimately involved in driving sexual desire. Love is the emotion of attachment and bonding to another person that is reinforced by oxytocin, a hormone synthesized in the hypothalamus and secreted into the blood by the pituitary. This cocktail of hormones coursing through the brain leads people to feel so strongly bonded to others that they are willing to die or kill for love.

More focused than a cancer diagnosis or a late-night death premonition is an execution date on death row. Between 1982 and 2016 the state of Texas executed 537 inmates, 425 of whom issued a last oral statement, which the Texas Department of Criminal Justice recorded and posted on its website, along with other details such as name, age, education level, prior occupation, prison record, and the offense for which they were executed.[27] This inadvertently created a database of the last thing these people (mostly men—only 7 of the 537 were women[28]) thought about just before they were put to death, on the gurney with the needles in their arms awaiting their lethal injections, in some cases as they were fading into unconsciousness narrating the end: "It's coming. I can feel it coming. Goodbye." And: "I feel it; I am going to sleep now. Goodnight, 1, 2 there it goes." Some were resigned to their fate, issuing brief expletive-filled declarations such as "Let's do it, man. Lock and load. Ain't life a [expletive deleted]?" and "I just want everyone to know that the prosecutor and Bill Scott are sorry sons of bitches." Other resignations were more dignified. "I'm an African warrior, born to breathe and born to die." But these were rare compared to the outpouring of love, sorrow, forgiveness, and blissful anticipation of the afterlife that is evident in a content analysis I conducted of all 425 final statements.

I became curious about these death row final sentiments after reading a 2016 examination of this dataset by the psychologists Sarah

Hirschmüller and Boris Egloff, who ran the statements through a computerized text analysis program called the Linguistic Inquiry and Word
Count (LIWC). The prisoners ranged widely in the number of emotional words they uttered, from 0 to 50 positive emotion words per
entry and from 0 to 27 negative emotion words per entry. To control for
this variation the psychologists computed an overall positivity index for
each death row inmate and found that 82.3 percent of them were above
0 in the use of positive emotion words. Comparing positive and negative emotion word usage, the biggest finding was a statistically significant difference between positive emotion words (9.64) and negative
emotion words (2.65).[29] Significant compared to what? To find out,
Hirschmüller and Egloff contrasted these findings with those reported
in another study of written words from a broad spectrum of sources,
including scientific articles, novels, blogs, and diaries, consisting of
over 168 million words composed by 23,173 people.[30] The mean of
2.74 positive emotion words for each entry in this dataset was statistically significantly lower than that of the prisoners (9.64). In fact,
these death row inmates were more positive than students asked to
contemplate their own death and write down their thoughts,[31] and
even more positive than people who attempted and/or completed suicides and left notes.[32]

This finding makes sense, given the fact that people on the verge of
committing suicide are in a different state of mind than those on death
row about to be executed. According to the psychologist Thomas Joiner,
in his book *Why People Die by Suicide*: "People desire death when two
fundamental needs are frustrated to the point of extinction; namely, the
need to belong with or connect to others, and the need to feel effective
with or to influence others."[33] Death row inmates, by contrast, used far
more social-orientation words, especially words referring to friends and
family.[34] After a decade or more on death row, these men develop relationships with other inmates and maintain connections with family and
friends on the outside, all of which obviate the motives characteristic of
those contemplating suicide.[35] Far from being terrified at the prospect
of their looming death, the outpourings of love in the Texas death row
inmate final statements supports Emotional Priority Theory over Terror
Management Theory.

To ensure that I did not cherry-pick examples in support of my thesis, I collaborated with my psychologist colleagues Anondah Saide and

Kevin McCaffree to enter all the statements into a database and then had two raters (Albert Ly and Liana Petraki) code each one based on preliminary categories I established after reading through all the statements myself, and a third rater (Marisa Montoya) reconcile any disagreements between the other two raters. From this we were able to compute inter-rater reliability correlations between coders that ranged from 0.50 to 0.83 and were all statistically significantly correlated at the 0.01 level of confidence. In other words, the coders consistently interpreted the statements in a manner significantly similar to one another and to my original analysis.[36]

Confirming my Emotional Priority Theory prediction, of the 425 death row inmates who made a statement, 68.2 percent used the word "love" (or a synonym for love) in reference to named girlfriends and wives, family and friends, and even to their fellow inmates. We excluded those indicating that they love God, Jesus, or Allah, which we included in a "religion" category. And although we did not count how many of them said they loved their mother, conspicuous by its absence is the fact that only one said he loved his father. I suspect (but cannot know for certain) that this is likely the result of so many of these men being raised in fatherless homes, a factor in the development of criminal behavior.[37] Figure 1-2 presents the results of our content analysis on the Texas death row inmates' final statements. I narrate each of these categories below.

Figure 1-2. Content Analysis of Texas Death Row Inmates' Final Statements

Of the 537 inmates executed by the state of Texas between 1982 and 2016, 425 issued a final statement. My colleagues Anondah Saide and Kevin McCaffree and I entered all statements into a database; I had two raters code each one based on preliminary categories I established after reading through all the statements myself, and a third rater reconcile any disagreements between the other two raters. The symbol k denotes the inter-rater reliability score between the coders and the $p < .01$ value signifies that the coders' ratings were statistically significantly correlated. The third figure for each category represents the percentage of statements made that included these expressed emotions and thoughts.

Love ($k = .832, p < .01$): 68.2%
Used the word "love" (or a synonym for love) in reference to family, friends, or other inmates (excluding those indicating that they love God, Jesus, or Allah).

Sorry (for Crime Committed) ($k = .790, p < .01$): 29.2%
Used the word "sorry" (or another synonym) in reference to committing the act or crime, but only if there was an admission of guilt. It was not counted if they were sorry for an "accident," or sorry that someone else committed the crime, or apologized to someone not involved with the crime (e.g., warden).

Forgiveness Requested ($k = .786, p < .01$): 14.1%
Asked for "forgiveness" from the family members of the victim(s), many apparently present to witness the execution.

Religion ($k = .831, p < .01$): 54.4%
Reference to or commentary about Jesus, God, Allah, Muhammad, or religion generally, not covered in the other categories.

Heaven or Afterlife ($k = .751, p < .01$): 33.6%
Reference to heaven, the afterlife, or another synonym referencing the hereafter.

Hell ($k = .496, p < .01$): 8.5%
Used the words "hell" or "evil" (or a synonym) in reference to the consequence of their crime.

Professed Innocence ($k = .842, p < .01$): 14.8%
Any claim about being innocent of their crime(s).

Capital Punishment Opinion ($k = .577, p < .01$): For: 2.8%; Against: 12.2%

Read the following excerpts from the death row inmates' final statements and ask yourself: Do these men sound as if they're in a state of terror, subconscious or otherwise? I think not. These statements read like expressions of emotions offered up as a final testament to what matters most to humans—love. Emotional Priority Theory better explains these sentiments:

> To my family, to my mom, I love you. God bless you, stay strong. I'm done.
> —Gustavo Garcia, February 16, 2016

I love you Renee, I am gonna carry your heart and always carry my heart in your heart. I am ready.

—Richard Masterson, January 20, 2016

I appreciate everybody for their love and support. You all keep strong, thank you for showing me love and teaching me how to love.

—Kevin Watts, October 16, 2008

I want to tell my sons I love them; I have always loved them—they were my greatest gift from God. I want to tell my witnesses, Tannie, Rebecca, Al, Leo, and Dr. Blackwell that I love all of you and I am thankful for your support.

—Hilton Crawford, July 2, 2003

As the ocean always returns to itself, love always returns to itself. So does consciousness, always returns to itself. And I do so with love on my lips.

—James Ronald Meanes, December 15, 1998

I would like to tell my son, daughter and wife that I love them.

—Jesse Jacobs, January 4, 1995

To my loved ones, I extend my undying love. To those close to me, know in your hearts I love you one and all.

—Ronald Clark O'Bryan, March 30, 1984

Following a preliminary discussion of my initial content analysis in one of my monthly *Scientific American* columns,[38] I received a letter from an artist and writer named Luis Camnitzer, who in 2008 produced an exhibition for the New York Gallery Alexander Gray Associates titled *Last Words*, featuring six human-sized prints in reddish-brown ink featuring some of these excerpts related to love.[39] The artist's intuitions about the importance of this emotion, so powerfully exhibited in the gallery (and partially reproduced in figure 1-3), is borne out by the data. Love matters, even to hardened criminals.

As a post hoc confirmation of Emotional Priority Theory, in 2016 a similar study was published on the last statements of 46 capital punishment inmates in the state of Missouri between 1995 and 2011, in which the researchers cataloged the statements into sixteen themes, the most

Figure 1-3. Luis Camnitzer's Exhibition *Last Words*

On human-sized sheets of paper in a 2008 exhibition at the Alexander Gray Associates gallery in New York City the conceptual artist Luis Camnitzer reproduced excerpts from the last words of the Texas death row inmates of their expressions of love. Photo courtesy of Alexander Gray Associates and Luis Camnitzer / Artists Rights Society (ARS).[40]

common of which was *love*, which garnered 54 percent. For example: "To my beloved children I want you to know I love you." "Tell my children and family and relatives, I love 'em." And "I can never express how much my wife means to me and how much I love her."[41]

In my own study, in addition to expressions of love, other emotional priorities evident in these final statements were admissions of *sorrow* for the crime committed (29.2 percent) and requests for *forgiveness* from the victims' families (14.1 percent).[42] Here is an archetypal example:

I'd like to apologize and ask forgiveness for any pain and suffering I have inflicted upon all of you, including my family. All of you, I am very sorry. There is a point where a man wants to die in judgment. Though my judgment is merciful, I hope and pray that all those involved as well as the judgment upon y'all, will one day be more merciful than mine. God bless you all. God speed. I love you. Remain strong. Ask God to have mercy. I love you all, too. I'm very sorry. I've got to go now. I love you.

—John Glenn Moody, January 5, 1999

Also evident was how much religious language appears in these statements. A majority (54.4 percent) indicated that they were religious, almost all Christian.[43] Here are a few typical statements:

> I thank God that he died for my sins on the cross, and I thank Him for saving my soul, so I will know when my body lays back in the grave, my soul goes to be with the Lord. Praise God. I hope whoever hears my voice tonight will turn to the Lord. I give my spirit back to Him. Praise the Lord. Praise Jesus. Hallelujah.
>
> —Hai Vuong, December 7, 1995

> Into your hands Oh Lord, I commend my spirit. Amen.
>
> —Peter Miniel, October 06, 2004

> I love you and I will see all of you in Heaven. I love you very much. Praise Jesus.
>
> —Troy Kunkle, January 25, 2005

> Jesus, thank you for your love and saving grace. Thank you for shedding your blood on Calvary for me. Thank you Jesus for the love you have shown me.
>
> —George Hopper, March 8, 2005

Given the power of such religious sentiments, it is not surprising that many of these men facing their death were not only *not terrified* at the prospect of death; they were looking forward to transitioning to the other side. Specifically, 33.6 percent of the statements included references to the afterlife in uplifting terms and phrases such as "going home," "going to a better place," going to the "other side," to "somewhere better," looking forward to "see[ing] each other again," "see you in eternity," "see you when you get there," "I'll be there waiting for you," "it's not the end, only the beginning," and of course references to heaven (but only 8.5 percent referenced hell). To wit:

> I know most of you are here to see me suffer and die but you're in for a big disappointment because today is a day of joy. Today is the day I'll be set free from all this pain and suffering. Today I'm going home to HEAVEN to

live for all eternity with my HEAVENLY FATHER JESUS CHRIST, and as I lay here taking my last breath, I'll be praying for all of you because you're here today with anger and hatred in your hearts letting Satan deceive you into believing that what you're doing is right and just.

—Clifton E. Belyeu, May 16, 1997

I just wanted to say to all of those that have supported me over the years that I appreciate it and I love you. And I just want to tell my mom that I love her and I will see her in Heaven.

—Demarco Markeith McCullum, November 9, 2004

Take care, give everybody my regards. I love you, and I'll see you in eternity. Father take me home. I am ready to go.

—Lonnie Johnson, July 24, 2007

This entry sums up all the emotional elements in one short dispatch:

To the West Family, I would just like to apologize for your loss. I hope that you can forgive me. To my family and loved ones and friends, I thank all of you all for your support and I am sorry for the pain and hurt I have caused you. I love you all and I will see you on the other side. O.K. Warden.

—Donald Aldrich, October 12, 2004

DEATH AND THE DEATH PENALTY: MORALISTIC PUNISHMENT AND MORAL CONSCIOUSNESS

Also revealing in our content analysis was the number of the men who said they were innocent or mistakenly convicted, set up by other criminals, wrongly accused by the police, or mistried by the courts, and were going to their deaths knowing that they didn't commit the crimes for which they were being executed. These were 14.8 percent of the total (not including the handful who said they were innocent because the murder they committed was an "accident"). For example:

I am innocent, innocent, innocent. Make no mistake about this; I owe society nothing. Continue the struggle for human rights, helping those who are innocent, especially Mr. Graham. I am an innocent man, and something very wrong is taking place tonight. May God bless you all. I am ready.

—Leonel Torres Herrera, May 12, 1993

I charge the people of the jury. Trial Judge, the Prosecutor that cheated to get this conviction. I charge each and every one of you with the murder of an innocent man. All the way to the CCA, Federal Court, 5th Circuit and Supreme Court. You will answer to your Maker when God has found out that you executed an innocent man. May God have mercy on you . . . Go ahead Warden, murder me. Jesus take me home.

—Roy Pippin, March 29, 2007

This brings me to the difficult topic of the death penalty in the context of thinking about the quest for human and social perfection that, given the fact that humans are not angels, necessarily requires a criminal justice system. A number of prisoners (15 percent) expressed their opinions on capital punishment, with 12.2 percent against it and 2.8 percent for it. Here is an example of an inmate statement in support of his execution:

My death began on August 2, 1991 and continued when I began to see the beautiful and innocent life that I had taken. I am so terribly sorry. I wish I could die more than once to tell you how sorry I am. I have said in interviews, if you want to hurt me and choke me, that's how terrible I felt before this crime. May God be with us all. May God have mercy on us all. I am ready. Please do not hate anybody because . . . [end of statement].

—Karl Chamberlain, June 11, 2008

Example of inmates' statements against the death penalty include:

I hope people understand the grave injustice by the state. There are 300 people on death row, and everyone is not a monster. Texas is carrying out a very inhumane and injustice. It's not right to kill anybody just because I killed your people. Everyone changes, right? Life is about experience and people change.

—Lee Taylor, June 16, 2011

This execution is not justice. This execution is an act of revenge! If this is justice, then justice is blind. Killing R.J. will not bring Anil back, it only justifies "an eye for an eye and a tooth for a tooth."

—Richard J. Wilkerson, August 31, 1993

What about the perspective of the families of the victims and their understandable desire for retributive justice? According to many of the death row inmates' final statements I read, prison was the hell from which death was a reprieve. Maybe life in prison is *worse* than execution. That's something to consider in the natural desire for justice.

Regardless of your beliefs about heaven and hell or your position on capital punishment, in this context it is clear that we all believe justice should be served in the here and now instead of (or in addition to) the hereafter. Perhaps this context explains the first and most famous Terror Management Theory study involving judges who were primed to think about their own mortality, after which they issued significantly harsher punishments than judges who were not so primed.[44] Maybe they do so not to reinforce cultural values as a means of attenuating personal death terror (the proposed explanation), but because humans have a deep-seated need to punish transgressors in order to maintain social harmony.

In his book *Moral Origins*,[45] the anthropologist Christopher Boehm argues that the emotion of *moralistic punishment* evolved in our Paleolithic ancestors to solve the problem of how relatively equitable hunter-gatherer societies remain stable when free riders could game the system by taking more than they put in. If everyone (or a majority of group members) cheated, lied, stole, or bullied, social harmony would disintegrate. To work around this problem, all fifty modern hunter-gatherer groups that Boehm studied employ sanctions to deal with deviants, free riders, and bullies, ranging from social pressure and criticism to shaming, ostracism, ejection, and even execution for unrepentant and irredeemable bullies. Of course, no justice system is one hundred percent efficient at preventing all violations, but in an evolutionary context, free riders and cheaters who respond to sanctions maintain their genetic fitness and pass on their genes for modest levels of free riding and cheating, which is what we see in all societies today. From this system we evolved a *moral conscience*, an "inner voice" of self-control.

In the context of Emotional Priority Theory, instead of the management of death terror, perhaps priming judges of their impending death triggered their moral conscience and reminded them to prioritize their sense of moralistic punishment, an emotion we all carry over from our evolutionary ancestry.

———

ONE OF THE most profound thoughts any of us can have is awareness of our own mortality, but it is not the primary driving force behind human thought and behavior, creativity and productivity. Our inability to imagine our own nonexistence means that an ultimate understanding of our own mortality will forever elude us, leaving us to live for the here and now even while the hereafter beckons us. Who were the first people to become aware of their own mortality, and to dream of immortality?

CHAPTER 2

WHAT DREAMS MAY COME

Imagining Immortality

Whether 'tis nobler in the mind to suffer
The slings and arrows of outrageous fortune,
Or to take arms against a sea of troubles,
And by opposing end them. To die; to sleep; . . .
To sleep? Perchance to dream! Ay, there's the rub;
For in that sleep of death what dreams may come.

—Shakespeare, *Hamlet*, act 3, scene 1

Have you ever seen someone die? I have. I was present when my mother took her final breath after succumbing to the wasting of a long coma induced by head trauma from a fall, itself the result of the ravages of a decade of brain tumors and the invasive treatment triad of surgery, radiation, and chemotherapy.[1] When the terminus was finally reached, after an initial interlude of loss and grief there followed a sense almost of relief. The anticipation was worse than the event. She was alive . . . and then she wasn't. What transpired at that moment? I don't know.

My father died in his car on his way to work—pulled over in a parking lot with the motor still running and the transmission in park. He must have known something was wrong. Was he aware he was dying? I have always wondered, and now that I'm older than he was when he died, I am curious if I will be given some warning—a thanatotic aura or the equivalent of a push notification from the grim reaper—before I go.

Will I experience a transition, see a light, go through a tunnel and emerge somewhere else, or will it be boom boom—lights out? When I recently went under general anesthesia for surgery, I imagined that perhaps this is what it's like to die. You're conscious, then you're not—99 . . . 98 . . . 97 . . . gone. But instead of coming to with missing time in between conscious states, you just never wake up. There's no missing time because time has stopped. Is that what death is like?

The descriptor "clinical death" gives us a clue that the dying process is sometimes more like a lamp dimmer than an on-or-off light switch. Some people die quickly, as my father apparently did from cardiac arrest, while others die gradually, as my mother did from coma wasting. The physician Sherwin Nuland penned an elegant book encompassing all aspects of the dying process titled, simply, *How We Die*.[2] The process is morbidly mesmerizing. *Clinical death* occurs when the heart stops beating and the lungs cease breathing. Without the oxygen-carrying red blood cells circulating throughout the body with each beat of the heart, deterioration of organs and cells begins. But that takes time. The oxygen that is already in the system lasts about four to six minutes following the final breath, which is why time is of the essence following cardiac arrest, drowning, and other traumatic events and why the CPR process of artificially pumping the heart and breathing into the lungs can be lifesaving. At room temperature, after around six to eight minutes—and no more than ten unless the body temperature has dropped dramatically, as in the rare cases of someone falling through ice into a frigid lake or river—*biological death* sets in as the rest of the body's organs shut down and cells die.

From this point forward the trillions of bacterial cells in the body that help digest food and provide other vital functions begin to consume the body's cells and tissues. After about an hour, *algor mortis*—death chill—sets in, as the body's temperature drops two degrees Celsius the first hour and one degree Celsius every hour thereafter until it reaches ambient room temperature. After around three to four hours, postmortem rigidity—*rigor mortis*—sets in, causing muscles to stiffen as a result of calcium bonding with the proteins in the muscles, leading to the unfortunate descriptor "the stiff" in reference to a dead body. Decay and disintegration are an exercise in clockwork chemistry from which forensic anthropologists and detectives can determine much about the time of death. As carbon dioxide levels rise, cell walls weaken and burst, releasing intercellular fluids that, as a result of gravity, pool into the nether regions of the body. Putrefaction results from the by-product of the foul-smelling gases including sulfur, ammonia, and sulfide produced by those trillions of bacterial cells in search of nutrients. Left alone, over the course of many months the body will be consumed biologically and chemically in a process that most of us would rather not contemplate and is probably why all societies everywhere and at all times bury their dead within days. In this light, in considering the archaeology of the afterlife among the earliest humans who appear to have buried their dead (as we will discuss below), in addition to speculating about their higher spiritual thoughts in so doing we should also entertain the possibility that they buried the dead bodies because they stank to high heaven.

So, we understand the physiological processes in the body and the neurological changes in the brain that accompany death, but comprehending what the "spark" of life is and where it goes when we die remains something of a mystery. We can't help but wonder what happens at that moment. We do not know, but at some point in our lives most of us ask such questions. At what age do we become aware of life's impermanence?

HOW CHILDREN CONCEIVE OF DEATH AND IMMORTALITY

The earliest memory I have of becoming aware of the finality of death was when my beloved dog Willy died. A scruffy-haired midsized mutt who was a bundle of playful energy for a young boy, Willy also afforded me much love and comfort at a particular time in my life when I needed it. My parents divorced when I was very young, and they both remarried, establishing two homes between which I was shuttled, and two new family configurations among which many adjustments had to be negotiated. My stepparents were warm and loving and as supportive of me as they were their biological issue, but it was a discombobulating experience nonetheless, so in quiet moments Willy gave me the kind of unconditional tenderness and devotion available only from canines. One day when I came home from school my mom told me Willy had died. As I recall, there was much sadness in the house, and I retreated to my bedroom to grieve as only a seven-year-old knows how: I cried and prayed for one more day with Willy. I felt bad for some time after, but then we got a new puppy, Kelly—an irresistibly adorable border collie whom I loved for his fifteen years on this planet.

The awareness of death's finality appears to emerge in preschool children around the age of four. Before that, children believe, for example, that dead animals can come back to life if given food, water, medicine, or magic potions. They see cartoon characters and TV actors die and come back to life, and they seem to conceive of the dead as living somewhere else, such as a tomb underground or heaven above, where they still consume food, water, and oxygen, can see and hear and dream, and continue to exist in some other state. This conception is reinforced by parents when grandparents pass away and the children are told that their senior family members have "gone to a better place" where they are still alive and maybe even "watching over them." (Maybe this is why I thought it possible that Willy might come back to me for a day.) Until they reach a certain age range children believe in immortality for everyone. Developmental psychologists tell us that that age range is between five and ten, when children come to recognize five features of death that make it real for them:

1. *Inevitability*. All living things eventually cease to live.
2. *Universality*. Death happens to all living things.
3. *Irreversibility*. Death is final, and once a living thing has died, it cannot come back to life.
4. *Nonfunctionality*. The bodily processes that characterize living things cease to function.
5. *Causation*. Death is the result of the breakdown of bodily function.

As with all stage theories in psychology, the timing and sequence of the stages vary, along with the ages at which children go through them, but the point is that by the age of ten, in the words of the clinical psychologists Virginia Slaughter and Maya Griffiths in their study on how young children understand death, "most children conceptualize death as a fundamentally biological event that inevitably happens to all living things and is ultimately caused by an irreversible breakdown in the functioning of the body."[3] That is, the dead cannot be resurrected. Slaughter and Griffiths encapsulate the chronology of how children reach this level of mortality awareness from infancy to age ten:

Infancy to age 2: The death of a parent or caretaker is felt as a loss, experienced as separation anxiety, and expressed in crying or changes in habits such as eating, sleeping, and activity, but there is no conception of death.

Ages 2–4: Preschoolers do not conceive of death as permanent and may wonder when the parent or grandparent who has died is returning. Their grief may be expressed as separation (and stranger) anxiety and they may exhibit higher than normal levels of clinging, bedwetting, thumb sucking, crying, temper tantrums, and possibly even withdrawal.

Ages 4–7: Death is still perceived to be reversible and superstitious causes are sought with an end to bringing back the deceased. Grief may be expressed in repetitive questioning, such as "What happens when you die?" and "How do dead people eat?" Eating and sleeping patterns change, possibly out of fear that they, too, may die.

Ages 7–10: The transition in the conception of death as temporary to permanent and reversible to irreversible takes place during these years. Children become curious about death and its causes, although they tend to see it as something that happens to old or sick people, and to people other than themselves or family members.

Ages 10–12: Conception of death is more intellectual than emotional, and grief may be expressed in silence, indifference, or regression from friends and family.

In three experimental tests of such conceptions with children of varying ages involving "the biological and psychological functioning of a dead agent," the psychologists Jesse Bering and David Bjorklund found that "4- to 6-year-olds stated that biological processes ceased at death, although this trend was more apparent among 6- to 8-year-olds," as predicted. In their second experiment, the researchers found that when 4- to 12-year-olds were asked about the *psychological* functioning of an agent, "the youngest children were equally likely to state that both cognitive and psychobiological states continued at death, whereas the oldest children were more likely to state that cognitive states continued." This is revealing because it implies that there may be cognitive architecture in the brain for the religious belief in an afterlife where the biological dead continue to live psychologically (or "spiritually" in religious terms). This hypothesis was tested and partially confirmed in the third experiment, in which both children and adults were asked about an array of psychological states. "With the exception of preschoolers, who did not differentiate most of the psychological states, older children and adults were likely to attribute epistemic, emotional, and desire states to dead agents."[4] And if you add to that list the "soul," then most adults hold such beliefs.[5]

Even very young children have a difficult time conceiving of the nonexistence of mind or soul. Using finger puppets representing a mouse and an alligator, Bering and Bjorklund presented a story to their young charges that the alligator ate the mouse, then asked them a series of questions: "Now that the mouse is no longer alive, will he ever need to drink water again?" "Is he still thirsty?" "Does his brain still work?" "Is he still thinking about Mr. Alligator?" Without exception the children maintained that the mouse's psychological states continued after the death of its body. The evidence overwhelmingly points to the thesis that belief in a psychological or spiritual afterlife is natural and intuitive, and that the scientific null hypothesis—that the afterlife does not exist and all psychological functions cease with biological death—is unnatural and counterintuitive. The psychologist Leslie Landon Matthews, whose famous father, the actor Michael Landon, died when she was thirty, poignantly illuminated the conceptual differences in this first-person account about her half brother and sister:

> Let us illustrate how the idea of reversibility affects different aged children. A father has died. Two months after dad dies, the four-year-old son, the seven-year-old daughter, and their mother go on a trip. They are away for one month. Upon their return, as they pull into the driveway, the four-year-old spots his father's car in the garage and yells out with excitement, "Daddy's home! Daddy's back home!" The seven-year-old has a split second feeling of excitement as well, but will quickly realize through her own higher level of understanding that her brother's comment is not true, and she may sob, for she knows that Daddy is not home.[6]

These bracketed age categories with their respective conceptualizations of death vary considerably within and between cultures and societies, but the point is that by our teenage years, we understand that death is inevitable, universal, and irreversible.[7] At the same time, most people also tend to believe that some part of life may continue into the next life, a tendency reinforced by most religions and the language parents use with their children to describe what happened to lost loved ones: they're "at rest," "at peace," "beyond the grave," "gone to a better place," "departed this world," are "in heaven with God," resting "at the feet of Jesus," "in Abraham's bosom," "in the Promised Land," and the like.

This sets up another mortality paradox: just as children come to understand the reality of death they are told that death is just a transition stage to some other place. This is a form of deceptive advertising we would not tolerate in other areas of life for our children. Revealingly, Slaughter and Griffiths ran an experiment with preschoolers in which they taught their charges biological facts about the body by having the kids don a cloth apron adorned with models of the body's organs and explained the organs' functions. They found that this led the kids to more rapidly develop a deeper understanding of those five characteristics of death (inevitability, universality, irreversibility, nonfunctionality, and causation). In a follow-up experiment with children aged four to eight, Slaughter and Griffiths discovered that the better the kids understood those five factors of death the less likely they were to express a fear of dying.[8] The cognitive psychologist Andrew Shtulman draws out the implications for the corrosive effects of confusing young minds about death in this way: "Children know of death long before they understand it, initially conceiving of death as an altered form of life. How frightening it must be for them to think that we bury people who still need food and water or that we cremate people who still think and feel pain. And how sad it must be for children to think that a loved one has left home but is still living on somewhere else."[9]

WHEN MAMMALS MOURN

Like us, dolphins are mammals, and they appear to pass the mirror test of self-awareness: put a giant mirror in their tank and a mark on their side and they'll stare at it as if it doesn't belong there, implying that they have some sense of body awareness, which is elemental to self-awareness.[10] It is not surprising, then, to hear stories from fishermen that dolphins have been seen pushing sick or wounded members of a pod to the surface so that they may catch their breath, and mothers who support their dead or dying offspring on their backs so that they may breathe. The marine biologist Filipe Alves recorded two such instances near Madeira Island off the northwest coast of Africa, in which the rescue dolphins made a concerted effort at resuscitation or resurrection.[11] Is this grieving behavior? Alves thinks so: "Species that live in a matrilineal system, such as killer whales and elephants; species that live in pods of related individuals, such as pilot whales whose pods can comprise up to four generations of animals, when they spend a lifetime together, sometimes 60 years or more, yes, I believe they can grieve."[12]

Observations of a bottlenose dolphin population in the Amvrakikos Gulf on the west coast of Greece by the biologist Joan Gonzalvo corroborates the behavior. A dolphin mother lifted the corpse of her newborn calf above the surface. "This was repeated over and over again, sometimes frantically, during two days of observation," Gonzalvo noted, adding, "The mother never separated from her calf. She seemed unable to accept the death." A year later the team came upon a juvenile dolphin struggling to stay afloat to breathe. His podmates were noticeably upset: "The group appeared stressed, swimming erratically. Adults were trying to help the dying animal stay afloat, but it kept sinking."[13]

Does this grieving behavior correspond to mortality awareness? We don't know for sure because we cannot know what it's like to be a dolphin, but marine biologist Ingrid Visser of the Orca Research Trust in Tutukaka, New Zealand, thinks that it is possible because "we do know that cetaceans have von Economo neurons, which have been associated with grief in humans." Whales also have these neurons, and when Visser observed a stranded pilot whale, she noted

> When one died the others would stop when passing by, as if to acknowledge or confirm that it was dead. If we tried to get them to move past without stopping, they would fight to go back to the dead animal. I do not know if they understand death but they do certainly appear to grieve—based on their behaviors.[14]

Whales have also been observed putting themselves in harm's way of whale hunters to protect or defend a wounded member of their group, and they have been seen circling their injured mate and striking the

water with their flukes, behavior that was exploited by the whale hunters to locate the targets of their hunt. What looks like self-destructive
behavior is in fact cooperative behavior tied to an apparent awareness
of the possible death of a podmate.

Elephants have also passed the self-awareness mirror test, and they,
too, appear to grieve. When they encounter the bones of long-dead elephants, especially the skull and tusks, they have been seen to stop and
ponder the find and carefully touch and move the bones with their trunks
in what looks like deep curiosity or concern. According to the animal
behaviorist Karen McComb, "their interest in the ivory and skulls of
their own species means that they would be highly likely to visit the
bones of relatives who die within their home range."[15] To test this hypothesis, McComb and her colleagues placed objects about 25 meters from
the elephants they were studying in Amboseli National Park in Kenya.
In the first condition, they planted the skulls of a rhinoceros, buffalo,
and elephant near seventeen different elephant families, noting that
their subjects spent the majority of their time carefully examining the
skulls of their own species, smelling and touching them with their trunks.
In a second condition a different set of nineteen elephant families were
confronted with a piece of wood, a piece of ivory, and an elephant skull.
Predictably, their interests scaled from most to least relevant: ivory,
skull, wood. But, McComb notes, "Their preference for ivory was very
marked, with ivory not only receiving excessive attention in comparison
with wood but also being selected significantly more than the elephant
skull." As McComb elaborated to me in an email, the elephants also
habitually touched and rolled the ivory with the sensitive soles of their
feet and picked it up in their trunks to hold and carry. Why? "Interest in
ivory may be enhanced because of its connection with living elephants,
individuals sometimes touching the ivory of others with their trunks
during social behavior." In the third condition, three elephant families
were presented with the skulls of three deceased matriarchs, one of
which was their own, but there was no observed difference in preference by the living elephants. That's not a trivial discovery, and McComb's
concluding remarks about the significance of the ivory touching is even
more revealing (figure 2-1): "Elephants may, through tactile or olfactory
cues, recognize tusks from individuals that they have been familiar with
in life." Imagine that: grieving over the remains of someone you knew.
How human.

In her moving memoir *Elephant Memories*, Cynthia Moss recorded the responses of a community of elephants to one of their members being shot by a poacher. As the wounded elephant's knees buckled and she began to go down, her elephant comrades struggled to keep her upright. "They worked their tusks under her back and under her head. At one point they succeeded in lifting her into a sitting position but her body flopped back down. Her family tried everything to rouse her, kicking and tusking her, and Tallulah even went off and collected a trunkful of grass and tried to stuff it into her mouth." After the elephant died, her friends and family members covered the corpse with dirt and branches.[16]

Hundreds of such anecdotes exist in scientific literature, and thousands more in popular prose.[17] There is, understandably, much skepticism of such accounts by more cautious scientists concerned about the anthropomorphizing of animals, but it is pertinent to note that we are animals, too, and just as there is an unmistakable continuity in our anatomy and physiology with our evolutionary cousins (about which

Figure 2-1. Elephants Grieving

The animal behaviorist Karen McComb photographed these elephants mourning the loss of family and group members. Photograph courtesy of Karen McComb.

no one accuses scientists of "anthropomorphizing"), so, too, with our behaviors and emotions, correspondences of which may be found, in some degree, in our fellow mammals, including and especially primates and cetaceans. These include not just base emotions such as *hunger, sex,* and *territoriality,* but more elevated emotions such as *attachment and bonding, cooperation and mutual aid, sympathy and empathy, direct and indirect reciprocity, altruism and reciprocal altruism, conflict resolution and peacemaking, deception and deception detection, community concern and caring about what others think about you,* and *awareness of and response to the social rules of the group.* The fact that such emotions exist in our nearest evolutionary cousins is a strong indication of their deep evolutionary roots. If we grieve over death, is it not reasonable to presume that these other closely related mammals do as well?

The grief psychologist Russell Friedman thinks so. Using his definition of grief as "the conflicting feelings caused by the end of or change in a familiar pattern of behavior," Friedman infers that since "all mammals are creatures of habit, there can be no doubt that mammals are affected by the deaths of their group members—if only because the deaths represent the end of the 'familiar' interactions for the surviving member." Thus, he concludes, "it is not far-fetched to suggest that the process of adapting might be affected by the nature and intensity of the individual relationship the surviving member had with the deceased member."[18] That resonates well with what we know about the evolution of nonhuman mammals. What about humans and our evolutionary past? When did our species first become aware that we are mortal?

SHADOWS OF OUR GRIEVING ANCESTORS

Thoughts don't fossilize, but sometimes actions do, such as burying the dead with flowers, belongings, or other people. For decades archaeologists thought that a burial site in the Shanidar Cave complex in the Zagros Mountains of northern Iraq might be the oldest example. Dated to around sixty thousand years ago, one of the sites contains the body of a man lying in a fetal position alongside two women and an infant child in a grave scattered with the pollen of flowers.[19] Indicative of the difficulty of fossil interpretation, however, archaeologists now think that the pollen was accidentally introduced into the site by the actions of animals, as nearby were found burrows of a gerbil-like rodent that is known to store seeds and flowers.[20] But how did the bodies end up in that configuration?

In 2013 it was announced in the *Proceedings of the National Academy of Sciences* that a fifty-thousand-year-old Neanderthal skeleton was intentionally buried at the famous La Chapelle-aux-Saints site in southwestern France. The bones were buried in a depression in the ground that archaeologists conclude could only have been intentionally dug, and taphonomic analysis of the fossils indicate that they did not show cracks and weathering, as was found in nearby bison and reindeer bones. "These multiple lines of evidence support the hypothesis of an intentional burial," the authors conclude.[21] Evidence from other Neanderthal sites indicates that individuals decorated themselves with pigments, wore jewelry constructed of colored shells and feathers,[22] and like the Shanidar finds, some showed signs of having been cared for by others after injury or in old age. One man, for example, was missing most of his teeth and had serious hip and back problems that may have required the assistance of others in order for him to survive (figure 2-2).[23] Keep in mind that Neanderthal brains were as large as our own, and though their cultural artifacts do not show the same rate of progress as those of early modern humans, they were sophisticated enough that we may reasonably infer that these were thinking and feeling hominids who had some awareness of their own mortality.

In *Homo sapiens* sites, features of ritual burial date back at least a hundred millennia. In modern Israel at the Skhul cave at Qafzeh, for example, archaeologists discovered hundred-thousand-year-old remains of a child buried with ritual goods and the antlers of a deer in its hands, several bodies in various contrived positions nearby, and the mandible of a wild boar grasped by the hands of another child.[24] A 2013 review study of eighty-five such burial sites ranging in age from ten thousand to thirty-five thousand years revealed that most were relatively plain with items from daily life, but a few contained more lavish grave goods such as ornaments of stone, teeth, and shells. Curiously, there was no sign of progression over time, as is often found in tools and other artifacts. "So, the behavior of humans does not always go from simple to complex," explained Julien Riel-Salvatore, the principal investigator of the study; "it often waxes and wanes in terms of its complexity depending on the conditions people live under." Also, intriguingly, the sites do not

Figure 2-2. Skull of Neanderthal Man from La Chapelle-aux-Saints Burial Cave
Photograph by DEA / A. Dagli Orti. Courtesy of Getty Images.

differ significantly from earlier Neanderthal graves, implying that the latter had the same cognitive capacities as modern humans, at least with regard to mortality awareness.[25]

A thirty-thousand- to thirty-four-thousand-year-old site at Sunghir, 120 miles north of Moscow, contains the remains of an adult man who was buried with 20 pendants, 25 rings, and 2,936 beads, all made of mammoth ivory and evidently sewn into his clothing (figure 2-3). Nearby is another grave containing a ten-year-old girl and a twelve-year-old boy who were also interred with some 10,000 ivory beads and other grave goods, such as mammoth tusk spears and hundreds of teeth of a species of arctic fox.[26] Similar grave goods were discovered in the Arene Candide cave on the Ligurian coast of Italy, dated to about twenty-nine thousand years ago, in which an adolescent male was laid to rest with hundreds of pierced deer canines and shells wrapped around his head, presumably originally sewn into a cloth or leather head covering, now disintegrated, along with mammoth ivory pendants, elk antler batons, and a ceremonial-length flint blade carefully placed in his right hand.[27] Such exquisitely carved items would have required considerable time to prepare, so whatever their purpose in life, it was evidently important to these Pleistocene hunters and foragers to equip their lost loved ones for the next life.[28]

More controversial still was the announcement in 2015 of a small-brained hominid species branded *Homo naledi,* whose fossil remains were discovered in the deep recesses of an almost inaccessible cave in South Africa. How did the bodies end up in such a remote location? The paleoanthropologists Paul Dirks, Lee R. Berger, and their colleagues suggested that the site represents the earliest example of "deliberate body disposal."[29] After the paper was published it wasn't long before "deliberate body disposal" was transmogrified into something more spiritually transcendent. Reuters, for example, announced: "Fossil first:

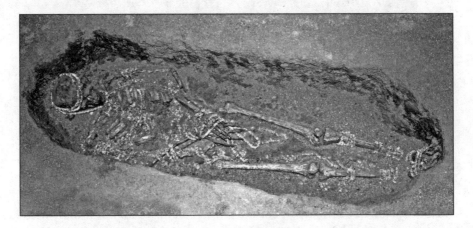

Figure 2-3. Burial of Man with Beads in Sunghir, Russia

Dated between 30,000 and 34,000 years old, this man was interred with 2,936 beads, along with 20 pendants and 25 rings, all made of mammoth ivory sewn into his clothing, since disintegrated, leaving this remarkable scene. Courtesy of José-Manuel Benito Álvarez.

ancient human relative may have buried its dead."[30] PBS inquired rhetorically: "Why did *Homo naledi* bury its dead?"[31] The discovery is controversial for a number of reasons, starting with the fact that the classification of the bones remains unclear as to where in the hominid lineage they fall, and their age is unknown. In a column in *Scientific American* I argued that intentional burial may be the result not of mourning but of murder,[32] but this hypothesis generated much skepticism, and most scientists are withholding judgment on the find until more research is conducted. Whatever the cause of death, however, and however long ago these hominids died, that such a small-brained primate would deliberately dispose of the bodies of their dead is a remarkable discovery and indicative of the deep evolutionary history of how our ancient ancestors dealt with death.[33]

What were any of these ancient hominids thinking when they buried their dead? Perhaps it was for purely sanitary reasons because, like so many other animals, they would deem it unhealthy to foul one's nest (or cave), so the prudent act would have been to bury the body. In addition, however, perhaps they also came to believe in something like what we conceive of as a soul. Did our ancient ancestors have some inchoate conception of an afterlife to which their charges would transcend from this life? We do not know, but at some point in those long-gone millennia the first beliefs in and conceptions about the afterlife arose. From there it was only a matter of time after the invention of writing some five thousand years ago that people began to compose stories and myths about the afterlife. These are the heavens of the world's major monotheistic religions—Judaism, Christianity, and Islam—to which we turn next along our journey.

CHAPTER 3

HEAVENS ABOVE

The Afterlives of the Monotheisms

And I saw a new heaven and a new earth: for the first heaven
and the first earth were passed away; and there was no more
sea. And God shall wipe away all tears from their eyes; and
there shall be no more death, neither sorrow, nor crying,
neither shall there be any more pain: for the former things are
passed away.

—Revelation 21:1, 4

Afterlife, afterworld, Arcadia, dreamland, Eden, Elysium, eternity, eternal
home, eternal rest, hereafter, higher place, holy place, kingdom come, land
of milk and honey, next world, nirvana, otherworld, paradise, Shangri-la,
wonderland, Zion. By any other name, *heaven* is the empyrean residence
of gods and other preternatural essences—angels, demons, ghosts,
souls—that have, to append a few common idioms, transcended,
crossed over, passed through, passed away, given up the ghost, or gone
the way of all flesh from the here and now into the hereafter. For millennia, scholars, theologians, religious philosophers, and men (and
women) of the cloth (and the book) have vested abundant time, energy,
and resources into making sense of this place where most people think
(and hope) they will go after death.

WHAT IS HEAVEN AND WHERE IS IT?

In the monotheistic religious traditions, heaven represents three broad conceptions: (1) a physical part of the universe, (2) the residence of God, and (3) a place to which the dead ascend. In Hebrew the word for "heaven"—*shamayim*—is sometimes rendered as "the firmament," as in "the heavens above." Cosmologically, the firmament is the dome-shaped canopy surrounding Earth that separated the heavenly waters above from the earthly waters below (whence came the water for the Noachian flood). The author of Genesis 1:6–8 explains it this way (King James Version):

> And God said, Let there be a firmament in the midst of the waters, and let it divide the waters from the waters. And God made the firmament, and divided the waters which were under the firmament from the waters which were above the firmament: and it was so. And God called the firmament Heaven.

This conception of heaven was adopted from the Ancient Near Eastern Mesopotamian cosmologies of the time, most notably Sumer, Babylon, and Judea between the eighth and sixth centuries B.C.E. (Before the Common Era) when the early books of the Old Testament were composed. The sixth century Priestly (or P) account that biblical scholars cite as one of the bases of the Torah (along with the Yahwist, Elohist, and Deuteronomist sources), for example, was most likely composed by Hebrew priests during the Jewish captivity in Mesopotamia. Cultural diffusion across the borders of contiguous political units was quite common in the ancient world. In addition to separating the earth from heaven, and the waters above from the waters below, the canopy was believed to be the medium in which the moon, sun, and stars were embedded (Genesis 1:14): "Let there be lights in the firmament of the heaven to divide the day from the night."

In broadest outline, the earliest Hebraic cosmology was a tripartite system with the heavens representing everything above the earth, the earth in the middle, and the underworld, or Sheol, below. By the fourth century B.C.E. the ancient Hebrews had inculcated Greek cosmology into their own, with a spherical earth surrounded by concentric spheres of heavens in which the stars and planets were embedded, with God outside the most distant sphere. As the centuries rolled by and scientific theories evolved along with instruments and observations, religious heavens evolved along with them, adjusting theological meanings accordingly. In this sense, the Bible is a wiki, edited for centuries by writers and scribes influenced by their surrounding cultures, until it was codified into an acceptable canon that was then reinterpreted according to the ideas of the cultures of the interpreters. Intelligent Design creationists today, who strain to fit the constantly changing discoveries and theories of modern astronomers into the Bible's cosmology, stand in a long tradition dating back three and a half millennia. In other words, the authors of the Old Testament borrowed ideas from other cultures and peoples whose cosmological theories were not much more sophisticated. Figure 3-1 shows what the cosmos looked like to these ancient peoples.

Imagine what the world looked like through ancient eyes by going outside on a cloudless night. Look up and visualize a crystalline dome over the earth in which all the stars are embedded. For the moon and planets that move against the background stars, picture each of them fixed in its own crystalline sphere rotating beneath, and independently of, the overarching dome. You are standing on a flat disk that is stationary and around which the spheres rotate. This is an intuitive world picture, because we don't feel the earth move. Our language reflects our intuitions, as when we speak of the sun setting or the stars rising. In this case, as in so many ideas about the world, our intuitions are wrong.

Nevertheless, this cosmological model of the heavens served a deeper role for the ancients involving heaven as God's residence and the repository of souls in an eternal afterlife, the ideas about which are as subject to cultural influences as those of the celestial heavens. Our focus here with reference to heaven will be on the monotheistic faiths of Judaism, Christianity, and Islam.

JEWISH BELIEFS ABOUT THE AFTERLIFE

Among adherents of the three monotheisms, Jews are the least likely to believe in immortality, with less than half of American Jews and only 56 percent of Israeli Jews professing belief in an afterlife and only 16 percent absolutely certain there is one (compared to the overwhelming majority of Christians and Muslims).[1] Tellingly for my

Figure 3-1. Biblical Cosmology

The Hebraic cosmology of Genesis was influenced by the Ancient Near Eastern Mesopotamian cosmologies of the eighth through the sixth centuries B.C.E., most probably those of Sumer, Babylon, and Judea. Illustration by Pat Linse.

thesis that heaven above and heavens on earth are often conflated historically, Jews are more focused on *this* world than the next, starting with the sacred covenant they made with God. As described in the Hebrew Bible, if the Jews followed God's commandments they would survive and flourish and be blessed with descendants and eventually be granted the land of Israel. The "world to come"—*Olam haba*—is a reference to the anticipated establishment of a just and fair society *here*, not in the *hereafter*.

The primary reference to some other world after death is Sheol, or an underworld lacking any characteristics of this world. This is not the hell of Christianity, but rather simply *nothing*. "For the living know that they shall die: but the dead know not any thing, neither have they any more a reward; for the memory of them is forgotten," reflects the author of Ecclesiastes (9:5). The long-suffering (and stupefyingly patient) Job pleaded with God (10:20–22): "Are not my days few? Cease then, and let me alone, that I may take comfort a little before I go whence I shall not return, even to the land of darkness and the shadow of death; a land of darkness, as darkness itself; and of the shadow of death, without any order, and where the light is as darkness." Sheol is different from Gehinnom, originally a place outside ancient Jerusalem where gentiles worshipped other gods, practiced child sacrifice, and possibly burned sacrificial corpses, which may be the origin of the myth of a fiery hell that later Christians transmogrified into a cosmic courthouse where sinners and wrongdoers were judged and punished, so gruesomely depicted by the fifteenth-century Dutch painter Hieronymus Bosch in his c. 1482 triptych *The Last Judgment*, particularly in the third panel's hellscape of eternal damnation of tortured souls (figure 3-2).

As for a heavenly afterlife, the book of Ezekiel, composed in the second century B.C.E., has God proclaiming (37:13), "And ye shall know that I am the LORD, when I have opened your graves, O my people, and brought you up out of your graves." By this rendering, the afterlife is achieved through the physical resurrection of the body, not the dislocation of the soul from the (forever) discarded body, an idea reinforced in the book of Daniel, also written in the second century B.C.E. (12:2): "And many of them that sleep in the dust of the earth shall awake, some to everlasting life, and some to shame and everlasting contempt."

By the twelfth century C.E. the revered Jewish philosopher Moses Maimonides confirmed that "The resurrection of the dead is the foun-

Figure 3-2. The Last Judgment

Hieronymus Bosch's triptych of *The Last Judgment*, depicting the Christian worldview. On the left panel are the Garden of Eden with God on his throne, good and evil angels in combat, the creation of Eve from Adam's rib, the serpent tempting Eve at the tree of the knowledge of good and evil, and Adam and Eve expelled from the Garden; the middle panel depicts the last judgment with Christ at the top accompanied by the Virgin Mary, John the evangelist, and the apostles, with the damned below them impaled, burned, hanged, and tortured; the right panel represents hell itself, where the damned are burned for all eternity.

dation of all grand principles of Moses, our teacher, and there is neither religion, nor adherence to the Jewish religion, in those who do not believe this."[2]

During the Middle Ages much debate took place about whether body and soul are inseparable, such that the death of one is the death of the other and the resurrection of one is the resurrection of the other. But by the Early Modern Period and into the eighteenth century, Jews began adopting the dualism of Descartes and other philosophers in believing that the soul alone survives death. Yet among the Dead Sea Scrolls discovered after the Second World War is a first-century C.E. text called the Damascus Document, which proclaims that God will "resurrect the dead . . . keeping faith with those asleep in the dust," implying physical resurrection of the body.[3] Either way—physical resurrection of the body or just the survival of the soul—the mechanism of immortality is less important than its possibility and what that means for morality and meaning, an emphasis reinforced by Christianity.

CHRISTIAN BELIEFS ABOUT THE AFTERLIFE

Christians believe that the resurrection of Jesus made him the messiah and, in turn, our savior and avenue to eternity, as pronounced in John 3:16: "For God so loved the world that he gave his only begotten Son, that whosoever believeth him shall not perish but have eternal life." Although many Christians take the physical resurrection of Jesus as evidence that humans, too, will be physically reincarnated in their original bodies in heaven, many other Christians hold that it is the soul alone that is eternal. The fact that Jesus was seen as a physical being after his death—with doubting Thomas memorably quelling his skepticism by inserting his fingers into Christ's chest wound—is for many Christians proof that physical resurrection is also possible for us, given that Jesus was fully human. Then again, Jesus is also believed to have been fully God, so the analogy for humans does not necessarily follow.

Christians have scoured the Bible for clues to the nature of what lives on into eternity. Job 19:25–26, for example, implies that when the body dies, a physical copy is resurrected in heaven: "For I know that my redeemer liveth, and that he shall stand at the latter day upon the earth: And though after my skin worms destroy this body, yet in my flesh shall I see God." As for who goes to heaven, according to the authoritative *Anchor Bible Dictionary* there are five biblical figures who reportedly ascended to the heavenly firmament: Enoch (Genesis 5:24), Elijah (2 Kings 2:1–12), Jesus (Luke 24:51), Paul (2 Corinthians 12:2–4), and John (Revelation 4:1). There are additional accounts of others alleged to have beheld the throne of God, which is presumably in heaven: Moses, Aaron, the elders of Israel, Micaiah, Isaiah, and Ezekiel.[4] Revealingly, the afterlife motif of the Old Testament parallels that of other Ancient Near East religions of the time, in which heaven was mostly the exclusive realm of the gods and not a place to which most people scaled after death, as in Psalm 115:

The heavens are the LORD's heavens,
But the earth he has given to the sons of men.

The dead do not praise the LORD,
Nor do any that go down into silence.

Going "down into silence" is another reference to Sheol, the place of nothingness from which no one returns. The Sheol we know as the "lake of fire" and "the Hadean world of the dead" doesn't appear until the last book of the Bible, Revelation, in which it is described as the place where the deceased are judged and either rewarded with resurrection in heaven or punished with eternal damnation in hell.

The ascent to an immortal heavenly life began with only extraordinary individuals making the journey (Moses, Aaron, Enoch, Elijah, Jesus, Paul, and others) but later became more inclusive, such that the souls of all people are eligible. New Testament writers adopted ideas from other cultures, particularly the concept that life on earth is temporary and provisional and that humans actually belong in heaven. "Gradually, in Jewish and Christian texts of the Hellenistic period, the older idea of the dead reposing in Sheol forever is replaced with either a notion of the resurrection of the dead or the immortality of the soul or some combination of the two," writes the biblical scholar James Tabor in the Anchor Bible, and the idea of "the righteous" being promised eternal life takes hold and was solidified with the death and resurrection of Jesus, after which that became the paradigm for "all righteous who follow."[5] Figure 3-3, *The Ladder of Divine Ascent*, a late-twelfth-century iconic painting of the monk John Climacus' work of the same title published in 600 C.E., depicts monks ascending the original stairway to heaven as Jesus welcomes them at the top.

Initially Christians believed that only a chosen few ascended to heaven, but later they came to accept that anyone could. This evolution parallels that of the ancient Egyptian god Osiris, who first appears in the pyramid texts around 2400 B.C.E. Osiris was said to be the giver of life in this world and the redeemer and merciful judge of the dead in the next world. Egyptian kings believed that as Osiris rose from the dead, so would they in union with him, thereby attaining eternal life for themselves alone. By the New Kingdom, everyone believed that if they accepted Osiris as their god, they, too, would be resurrected from the dead. Democratizing the afterlife in this way is an effective tool for recruiting more members, and it has served many a religion well.

Figure 3-3. The Ladder of Divine Ascent

Painted in the twelfth century, this image depicts the thirty rungs of the ladder representing the thirty stages of the ascetic life; the demons grappling the monks represent the many temptations of this life that might prevent one from reaching the next life with God and Jesus. Jesus, upper right, welcomes monks who made it, while angels upper left and monks bottom right encourage seekers to press on. At the bottom left, Satan devours a fallen monk. From the collection of Saint Catherine's Monastery, Mount Sinai.

In the twelfth century the Catholic Church added the doctrine of *purgatory*, a type of spiritual spa where people go to be purified before entering the Kingdom of Heaven. This led to the sale of indulgences— prayers you could purchase for your recently deceased loved ones in order to expedite the purging process before they made the celestial journey. Martin Luther famously broke with the Vatican over this and other abuses, leading to the Protestant Reformation and the catastrophic European wars of religion. Purgatory should not be confused with *limbo*, a thirteenth-century concept that Pope Pius X canonized in his 1905 catechism as a place for infants to go who died before baptism, along with all the Old Testament patriarchs who lived before Jesus, for obvious reasons.

The ever-evolving nature of religious beliefs was on display in 1999 when Pope John Paul II determined that heaven and hell are not actual physical places but states of the soul in communion (or not) with God: "The 'heaven' or 'happiness' in which we will find ourselves is neither an abstraction nor a physical place in the clouds, but a living, personal relationship with the Holy Trinity. It is our meeting with the Father which takes place in the risen Christ through the communion of the Holy Spirit."[6] Hell is not other people (as Jean-Paul Sartre famously opined in *No Exit*), but separation from God. Protestants, not always on board with what Catholics declare as truth (how did the pope determine that heaven and hell are not real places, anyway, beyond the usual armchair ratiocination?), rejected that interpretation, sticking to the dogma that heaven and hell are real and that people should look forward to the former and fear the latter.

Once you get to the Christian heaven, what's it like? Since no one has ever gone and come back with irrefutable evidence, believers must once again be content with biblical or theological narratives, sprung entirely from the imagination of the narrators. Some saints claim to have had visions of heaven, but visions are just another form of fiction and are not reliable. Revelation 22:5 says that in heaven "there shall be no night there; and they need no candle, neither light of the sun; for the Lord God giveth them light: and they shall reign for ever and ever." The book of Revelation is filled with visionary descriptions of heaven by its author, John, including (Revelation 4) "four and twenty elders sitting, clothed in white raiment; and they had on their heads crowns of gold," "seven lamps of fire burning before the throne," "a sea of glass like unto crystal: and in the midst of the throne, and round about the throne, were four beasts full of eyes before and behind. And the first beast was like a lion, and the second beast like a calf, and the third beast had a face as a man, and the fourth beast was like a flying eagle. And the four beasts had each of them six wings about him; and they were full of eyes within." According to John, heaven appears to be a duplicate of earth and its inhabitants, but without the downside: "And God shall wipe away all tears from their eyes; and there shall be no more death, neither sorrow, nor crying, neither shall there be any more pain: for the former things are passed away" (21:4). The book goes on and on in this vein. Seals and horses and eclipses and earthquakes and fire and hail and beasts and angels and dragons and seas of blood . . .

Broadly speaking, Christians believe that they will live for all eternity with God. What this means, exactly, is not clear. Is it a static and unchanging experience, or do people continue to grow and learn? Does heaven already exist, or is it a future state after the second coming of Christ? If it is a physical place, is it three-dimensional or some other dimension? Recall that the Hebrew word for heaven, *shamayim*, is plural ("the heavens"), so multiple dimensions seem as plausible as multiple places, or even one place with many rooms. Perhaps heaven is the pure omnipresence of God as described in Jeremiah 23:24: "Do not I fill heaven and earth? says the Lord." But there also appears to be a higher heaven still, as in Ephesians 4:10, in which Jesus "ascended far above all the heavens." Above the heavens?

Those less concerned with what a literal heaven is like are more inclined to employ the Hebrew Gan Eden, or the Garden of Eden, a more metaphorical place where people lived in perfect harmony before the fall and the invention of sin. A paradisiacal garden is what you might expect from the reveries of a desert-dwelling people, where fresh water, ripe fruit, abundant crops, lush vegetation, and dense herds of eatable domesticated ungulates, along with milk and honey and oil and wine, were in short supply. The word "paradise," in fact, is derived from *pairidaeza*, or "walled garden," and the envisioned New Jerusalem is a walled city, although this does lead one to wonder, what are the walls for in a perfect paradisiacal world? Could it be that heaven is to be found only on earth, that it is not perfect or paradisiacal and thus needs protective bulwarks?

ISLAMIC BELIEFS ABOUT THE AFTERLIFE

Although Allah is said to be the same God as that of the Jews and Christians,[7] the differences in the religions begin with the meaning of the word Islam—*submission*. On the Day of Judgment the fate of a Muslim's soul is assessed by a standard of submission to the three central tenets of the faith: (1) the authority and perfection of the Qur'an, (2) the monotheistic belief in God (Allah), and (3) the recognition of Muhammad as a prophet of God, as in the Islamic creed "There is no god but God and Muhammad is his prophet."[8] What happens then? According to the Qur'an (22:7): "Truly the Hour is coming—there is no doubt of it—when God will resurrect those who are in the graves." After the physical resurrection of the dead, judgment then determines their fate in the now familiar binary model of heaven and hell. "In the next world, paradise is the human realm, while hell is the realm of those creatures who began as human beings but did not live up to their humanity," write the religious scholars Sachiko Murata and William Chittick in their book *Visions of Islam*. "In Islam, paradise is always juxtaposed with hell, while heaven is always contrasted with earth. Paradise and hell pertain to the Return to God. They cannot be experienced in their fullness until after the Last Day."[9] Like the Bible, the Qur'an does not date the "hour of doom," but it does describe what it will be like when the end comes (81:1–14):

> When the sun shall be darkened,
> When the stars shall be thrown down,
> When the mountains shall be set moving,
> When the pregnant camels shall be neglected,
> When the savage beasts shall be mustered,
> When the seas shall be set boiling, . . .
> When hell shall be set blazing,
> When paradise shall be brought nigh,
> Then shall a soul know what it has produced.

Just as the Hebraic worldview was influenced by the cosmological models of other Near Eastern peoples, the scribes who collected and edited the texts that make up the Qur'an adopted the cosmology of seventh-century Near Eastern cultures, referring to seven heavens that correspond to the seven "planets" visible to the naked eye (Mercury,

Venus, Mars, Jupiter, and Saturn, plus the sun and moon). In fact, it was to the heaven of the moon that the angel Gabriel took Muhammad by the hand to meet Adam, and then to the successive six heavens where they met other prophets, including Abraham, Moses, and Jesus. They also visited paradise (Janna) and hell (Jahannam). Which one a Muslim goes to after death depends on the choices made in life. There are right deeds and right beliefs, and evil deeds and evil beliefs. "The good deeds of each person will be put in one pan and the evil deeds in the other. If good deeds predominate, the person will go to paradise, but if evil deeds predominate, he or she will be thrown into hell." The judgment instrument (*mizan*, or balance) appears to be something akin to the scales of the Greek goddess Justitia: "The scrolls will then be put on one side of the scale, and the document on the other, and the scrolls will become light and the document heavy, for nothing can compare in weight with God's name," according to one of the *hadiths* (accounts) recognized by Muslims as second only to the Qur'an in scriptural sanctity.[10]

At death the body remains in the grave while the Angel of Death takes the soul for judgment. Until resurrection, the deceased are in a state of separation (*barzakh*), during which they can experience a life review and anticipate a trip either to paradise or to hell. Two angels named Munkar and Nakir visit the deceased and ask "Whom have you worshipped?" and "Who was your prophet?" The correct answer—"There is no god but God, and Muhammad is His prophet"—gets you to Judgment Day in a roomy grave from which you may glimpse paradise. An incorrect answer—which is pretty much anything that veers off the Qur'anic course—will send you to a confined grave where you will be attacked by scorpions and spiders while looking out the window onto hell, which is predictably hot because of the fires, with boiling water and food that fails to nourish, if it doesn't choke you first. "Every living person shall taste of death, and we test you by evil and by good as a trial," God says in the Qur'an. "Whosoever has done even an atom's weight of good will behold it; and whosoever has done even an atom's weight of evil will behold that." Unlike Christianity, in Islam there is no concept of "original sin," so there is no need of a savior to die for your sins. Thus, Muslims do not recognize Jesus as the messiah.

As with other desert-dwelling peoples, Muslim scripture describes paradise as a garden that includes flowing water, along with milk, wine, honey, dates, pomegranates, and other earthly delights one might crave with no supermarkets in sight. It is an understandable desire given their environment. Mecca, for example, was described by a tenth-century Arab geographer named Maqdisi as "suffocating heat, deadly winds, clouds of flies," a place so insufferably hot that the people described the city as "burning."[11] By contrast, this garden paradise consists of "green, green pastures," "cool pavilions," and "fountains of gushing water." There, according to the Qur'an, the resurrected bodies of the Muslims judged worthy are adorned in silk clothes and gold jewelry; they are housed in beautiful mansions and provisioned with "fruits of every kind, and they will be honored in gardens of delight, sitting on couches face-to-face, with cups from a flowing stream being passed around, clear, delicious to drink, neither dulling the senses nor intoxicating." Naturally there's sex in paradise. If you were married in this life you continue to have sexual relations with your spouse in the next life . . . for eternity (cue comedic riffs on marital sex here). If you were single in this world you get hooked up with beautiful companions in the next, by some accounts no fewer than seventy-two, all virgins, naturally. (Just in case, according to one *hadith*, extra sexual potency is an added bonus for those who need it, presumably a form of celestial Viagra.) As we saw in the prologue, this fantasy has been used as a major motivator for armed jihad, whose warriors are literally dying to go to heaven.

HEAVEN AND ITS DISCONTENTS

An immersion in the extensive scholarly literature on the subject draws one to an inescapable observation: heaven has a history. J. Edward Wright's *The Early History of Heaven*,[12] Alister McGrath's *A Brief History of Heaven*,[13] Jeffrey Burton Russell's *A History of Heaven*,[14] and most notably Alan Segal's authoritative *Life After Death: A History of the Afterlife in Western Religion*[15] and Colleen McDannell and Bernhard Lang's definitive *Heaven: A History*[16] reveal what a fluid and historically bound concept heaven is.[17] "Imagining a heaven," writes Alan Segal in summarizing his exhaustive historical review, "involves projecting our own hopes on heaven and then spending our lives trying to live up to them."[18] "In learned tomes of theology in sermons, letters of consolation, poetry, the visual arts and countless unrecorded conversations," McDannell and Lang reflect on their millennia-long journey

into the past, "visionaries claim to have travelled to the beyond; philosophers present their reasoned speculations; artists paint their inner
visions. Their descriptions show a remarkable variety." The variation in
heavenly themata, in fact, is staggering. "For some, life everlasting will
be spent on a 'glorified' earth. Others think of heaven as a realm outside
of the universe as we know it. There are those who predict an eternal
life focused exclusively on God. Still others describe individual friendship and marriage. Eternal rest vies with eternal service." What are we to
make of this diversity of ideas? McDannell and Lang conclude:

> There is no basic Christian teaching but an unlimited amount of specu-
> lation. For the theologian, the lack of agreement on what goes on in
> heaven may be disappointing. For the philosopher, the idea that heav-
> enly doctrines provide no fixed ontological structures can be frustrat-
> ing. For the historian, however, such change and variety is a delight.[19]

As a historian of science, I, too, find this diversity a delectation, but
for the scientist such variation of beliefs is indicative of the likelihood
that none of them are "true" in any ontological sense. It is not just that
they are so obviously culturally bound and geographically determined,
but also that there are no means of determining which ones are more or
less likely to match reality. Cosmology has a history, for example, but
histories of cosmology differ from those of heaven in one important
way: there really is a cosmos, and our understanding of it really has pro-
gressed toward a better understanding as a result of better data and
more sophisticated theories. There is nothing comparable in the history
of heaven, no dates at which important discoveries were made that
illuminated our understanding of the true nature of the afterlife.

Among its many manifestations in history and literature, heaven is
not only where God is; it is the site of a cosmic courthouse where ultimate justice will be meted out to reward people for their good works in
this life and, by exclusion, to punish sinners. It is a repository not only
of disembodied souls but, for some Christian sects, of physical bodies
resurrected to their ideal age (thirty-three by some accounts, because that
was the age of Jesus when he was crucified) and perfect health (the blind
will see, the deaf will hear, the crippled will walk). It is a storehouse for
spiritual bodies or souls, resurrected with all the memories of the physical body but without the container. As pollsters have discovered, most

people conceive of heaven as a place where there will be no trials and tribulations, not a care or worry, no sickness or pain, no sorrow or pity. The afterlife will be happy and joyful, blissful and peaceful, and filled with love.[20] Heaven is said to be the opposite of hell, and as in the Hotel California, you can check out any time you like but you can never leave. Heaven is the final destination at the terminus of one's life, the completion of earthly history, the end of days.

The heavenly visions of Jews, Christians, and Muslims, in fact, are exactly what one might expect from ancient desert-dwelling peoples setting up a 1.0 operating system for answering life's hardest questions and for controlling the masses with a one-dimensional binary carrot-and-stick moral scheme. Imagine trying to operate a computer thousands of years from now using Windows 98. Software programs must evolve with hardware equipment. The modern world of science and reason, and the corresponding secular values of rights and justice, have about as much in common with these ancient religions as a modern computer with Internet access has with Charles Babbage's nineteenth-century Analytical Engine with its wheels, gears, and paper punch cards. Science and technology progress, and so should morals.

Thanks to the bending of the arc of the moral universe the sphere of inclusiveness has expanded in all areas of moral consideration, including among Christians on the matter of who gains admission to heaven. For much of Christianity's history, *exclusivists* held sway, believing that only Christians are eligible to be saved; but more and more, *inclusivists* are gaining ground, holding that people of different faiths may be eligible for heaven. In the early twentieth century, for example, fewer than 10 percent of Christians were inclusivists in their belief about who might go to heaven, but a 2008 Pew Forum survey found that 57 percent of Evangelicals, 79 percent of Catholics, and 83 percent of Protestants agreed that "many religions can lead to eternal life," while only 29 percent of Christians in general said "my religion is the one, true faith leading to eternal life." Tellingly, the survey found a positive correlation between church attendance and exclusivity, with regular churchgoers more than twice as likely as those who attend religious services less often to believe theirs is the one true religion leading to eternal life, the effect most notable among Evangelicals.[21]

Different religions have different versions of heaven and the afterlife (and how to get there). Egyptians imagined a physical place far above the earth in a "dark area" of space where there were no stars. The Vikings fanatisized about Valhalla, which for some included a big hall where they would drink beer and get ready to fight again. Muslims envision a garden with rivers, fountains, shady valleys, trees, milk, honey, and wine—all the things a desert-dwelling people would crave. And Christians, of course, picture an eternity with angels at the throne of God. Which one is right? By what criteria is one to assess the competing and conflicting heavenly hypotheses? When scientists are confronted with opposing hypotheses, they run experiments or compare datasets to see which one is most likely to be true, or least likely to be false. Theologians have no such tools at their disposal. As the preeminent medieval Jewish philosopher and astronomer Moses Maimonides wrote in the Mishneh Torah (1170–1180 C.E.): "As to the blissful state of the soul in the World to Come, there is no way on earth in which we can comprehend or know it."[22]

Even the desirability of heaven is no certainty. When the *Saturday Night Live* comedian Julia Sweeney was told by Mormon missionary boys that in heaven her body would be returned to its original state, she wondered, "What if you had a nose job . . . and you *liked* it? Do you have to get your old nose back?" After she explained to her interlocutors that she had her cancerous uterus removed and they told her she would get it back, she told them, "I don't *want* it back!"[23] The nineteenth-century ethnologist Élie Reclus described the resistance Christian missionaries faced when attempting to convert Inuits with the promise of a Christian-like heaven:

> Inuit: And the seals? You say nothing about the seals. Have you no seals in your heaven?
>
> Missionaries: Seals? Certainly not. We have angels and archangels . . . the 12 apostles and 24 elders, we have . . .
>
> Inuit: That's enough. Your heaven has no seals, and a heaven without seals is not for us![24]

My first college professor, Richard Hardison, had this question about heaven: "Are there tennis courts and golf courses?" In other words, are there any challenges? If there is no disease, sickness, aging, or death in heaven, if there are no obstacles to overcome and nothing to work for, what is there to do? Forever is a long time to be blissfully bored. If the

Christian version of heaven is correct and you get to spend eternity with an omniscient and omnipotent deity who knows and controls everything you think, do, and say, then as Christopher Hitchens famously opined, that would make heaven a "celestial North Korea" from which "you would never be able to escape,"[25] a "place of endless praise and adoration, limitless abnegation and abjection of self."[26] As Hitch told a UCLA audience in one of his last performances on stage shortly before his death in 2011:

> It will happen to all of us that at some point you get tapped on the shoulder and told not just that the party's over, but that the party is going on and you have to leave. That's the reflection I think that most upsets people about their demise. Alright, then, let's pretend the opposite. Instead you'll get tapped on the shoulder and told there's great news. This party's going on forever and you can't leave![27]

Even the concept of immortality is ultimately incomprehensible to us mortal beings, in the same manner that infinity is unfathomable to a finite being. What could it possibly mean to exist for an eternity? As Woody Allen said, "eternity is an awful long time, especially toward the end."[28] Perhaps this is why religious depictions of paradise resemble earthly scenes, only without such negative elements as darkness, predators, hunger, pain, exhausting work, and suffering. Instead, abundant resources and ease of life are vouchsafed to all who make it to these heavens. The source of such paradisiacal visions is so obviously terrestrial that it strongly suggests that it is entirely contrived by people suffering from the hardships wrought by daily life. Consider these famous passages from the book of Isaiah (65:17–25 passim):

> For behold, I create new heavens and a new earth; and the former things shall not be remembered or come into mind.
>
> No more shall there be in it an infant that lives but a few days, or an old man who does not fill out his days, for the child shall die a hundred years old, and the sinner a hundred years old shall be accursed.
>
> They shall build houses and inhabit them; they shall plant vineyards and eat their fruit.
>
> They shall not labor in vain, or bear children for calamity; for they shall be the offspring of the blessed of the LORD, and their children with them.

The wolf and the lamb shall feed together, the lion shall eat straw like the ox; and dust shall be the serpent's food.

They shall not hurt or destroy in all my holy mountain, says the LORD.

Is this a description of a heavenly paradise, or an earthly one? According to the authoritative Interpreter's Bible, in these passages "the meaning is not that the present world will be completely destroyed and a new world created, but rather that the present world will be completely transformed . . . there is no cosmological speculation here."[29] Indeed, in the Hebrew Bible it is not until the book of Daniel—the latest addition to the canon—that one can find reference to humans ascending to heaven. For these ancient believers, heaven was on earth, not a place to go to after life.

The shift from earthly paradise to cosmological firmament began in the book of Daniel and was reinforced in the New Testament, especially by Jesus, who suggested to his oppressed peoples that redemption was just around the chronological corner. Yet even Jesus made intriguing references to the kingdom that "has come upon you" (Luke 11:20), and especially in Luke 17:20–21, where he seems to imply that heaven is a state of mind: "And when he was demanded of the Pharisees, when the kingdom of God should come, he answered them and said, The kingdom of God cometh not with observation: Neither shall they say, Lo here! or, lo there! for, behold, the kingdom of God is within you."

Perhaps this interpretation illuminates the tantalizing passage in Matthew 16:26 in which Jesus told his disciples, "Verily I say unto you, There be some standing here, which shall not taste of death, till they see the Son of man coming in his kingdom." The passage has long been quoted by skeptics in response to the Christian claim that the end is nigh, that the second coming is upon us, and that Jesus will return at any moment. Maybe Christians have been misreading such passages for centuries. Maybe the "kingdom" to which Jesus refers is the heaven within ourselves, and the heavenly communities we build here on earth. Heaven is not a paradisiacal state in the next world, but a better life in this world. Heaven is not a place to go to but a way to be. Here. Now. Since no one—not even the devoutly religious—knows for certain what happens after we die, Jews, Christians, and Muslims might as well work toward creating heavens on earth.

PART II

THE

SCIENTIFIC SEARCH

FOR IMMORTALITY

To live on a day-to-day basis is insufficient for human beings; we need to transcend, transport, escape; we need meaning, understanding, and explanation; we need to see over-all patterns in our lives. We need hope, the sense of a future. And we need freedom (or, at least, the illusion of freedom) to get beyond ourselves, whether with telescopes and microscopes and our ever-burgeoning technology, or in states of mind that allow us to travel to other worlds, to rise above our immediate surroundings.

—Oliver Sacks, "Altered States," *The New Yorker,* 2012

CHAPTER 4

HEAVENS WITHIN

The Afterlives of the Spiritual Seekers

I died as a mineral and became a plant,
I died as plant and rose to animal,
I died as animal and I was Man.
Why should I fear? When was I less by dying?

—Rumi, *The Ascending Soul*[1]

My dosha is a Pitta. So I was told by the medical director of the Chopra Center, an inviting complex located at the posh La Costa Resort in the coastal town of Carlsbad, California. I went there in February 2016 to experience firsthand the world and worldview of the spiritual seekers, particularly as they are practiced by their most prominent American proponent, Deepak Chopra. Doctor, author, speaker, meditator, and practitioner of complementary and alternative medicine, Chopra is arguably the most prominent figure in the New Age movement today, having served as a spiritual guru to millions, including celebrities such as Michael Jackson and Oprah. Many people have turned to these spiritual traditions in search of something they can't seem to find in Western religions or science.

That same month, for example, I attended an event hosted by Deepak and another New Age spiritual seeker named Eckhart Tolle—best known for his massively popular book *The Power of Now*—in which the duo packed the sixty-three-hundred-seat Shrine Auditorium in

Los Angeles with admiring followers and Hollywood luminaries. For an hour and a half the two sages extolled the virtues of meditation, conscious awareness, and living in the present—the *now*. Cognitive psychologists have determined that "now" represents roughly three seconds of awareness. In Tolle's worldview, everything that came before *now* is past, which you cannot change, and everything that comes after *now* is future, which has yet to happen. *Now* is the only thing you can experience and so this is where (and when) your power lies.

I have only the vaguest notion of what any of this means. I once spent a long weekend at the Esalen Institute in Big Sur, California, a retreat center devoted to meditation, massage, yoga, personal growth, organic food, and clothing-optional natural hot springs. There I managed to live in the *now* from Friday afternoon through Sunday evening. But then I had to go back to work on Monday morning because my mortgage payment's *now* was coming up *soon*. Weekends are good for living in the *now*, weekdays not so much. That Western way of structuring life is what these spiritual-seeking sages want to change because heaven is within us, not above us.

My attempt to understand Deepak Chopra's worldview has been under way since the mid-1990s, when we published a cover story on him in *Skeptic*. Our relationship has been on a roller coaster ever since, alternating between respectful exchanges and dismissive snarls. We have debated science, religion, God, and the afterlife at conferences, on television shows, and privately over meals. I have long been critical of his worldview and he of mine. I have charged him with practicing pseudoscience and for spouting "woo-woo" nonsense, and he has reproached me for narrow-mindedness, dogmatic materialism, and excessive scientism. We were at an impasse in our relationship, so to bridge the gap and to help me better understand his worldview Deepak invited my wife and me to attend a three-day retreat at the Chopra Center, integrating massage, yoga, and meditation, with a strong dose of Eastern philosophy and Vedic science. In this chapter I will address ideas about the afterlife and immortality from the perspective of these spiritual traditions, particularly because they are aligned with what scientists are working toward in achieving immortality that we will consider in subsequent chapters.

THE WAR OF THE WORLDVIEWS: DUALISM VS. MONISM

One fundamental distinction to make in worldviews is that between *dualism* and *monism*. *Dualists* believe that we consist of two substances—body and soul, brain and mind (called "substance dualism" by philosophers). *Monists* contend that there is just one substance—a body and a brain—from which consciousness is an emergent property, "mind" is just the term we use to describe what the brain is doing, and the soul is just the pattern of information that represents our thoughts, memories, and personalities. As such, monists hold that the death of the body—the disintegration of the material body and the degradation of memory patterns in the brain—means the death of the soul. By contrast, dualists assert that the soul, like the mind, is a separate entity from the body, so even after the death of the body the soul continues.

Most people are dualists because dualism is intuitive—it just feels as if there is something else inside us, in the same way that the thoughts floating around up there in our skulls feel like a mind separate from our brain. The psychologist Paul Bloom calls us "natural-born dualists," as reflected in our language when we use phrases such as "my body aches" (rather than "I ache") or "my mind is muddled" (rather than "I am muddled"), as if "I" and "body" and "mind" are separate things."[2] In his delightful chapter on "Homer's Soul" in *The Psychology of the Simpsons*, Bloom shows how pervasive this dualism is in pop culture, as in the episode in which Homer is trying to figure out why his phone bill is so high:

> *Homer*: Burkina Faso? Disputed Zone? Who called all these weird places?
>
> *Homer's Brain*: Quiet, it might be you! I can't remember.
>
> *Homer*: Naw, I'm going to ask Marge.
>
> *Homer's Brain*: No, no! Why embarrass us both? Just write a check and I'll release some more endorphins.
>
> [Homer scribbles a check, then sighs with pleasure.]

The scene is humorous because as intuitive dualists we get the joke, but as monists we know that there is no dualistic split between Homer and Homer's brain. There's just Homer's brain talking to itself.[3]

In his lab testing young children's cognitive development, Bloom and his team tell kids a story about a human brain that has been transplanted into the head of a pig. As natural-born dualists they think the animal will still act like a pig, with the same personality and memories of a pig, only smarter. Bloom extrapolates such experimental findings to draw inferences about the development of the dualistic belief in immortality:

> This is the foundation for the more articulated view of the afterlife you usually find in older children and adults. Once children learn that the brain is involved in thinking, they don't take it as showing that the brain is the source of mental life; they don't become materialists. Rather, they interpret "thinking" in a narrow sense, and conclude that the brain is a cognitive prosthesis, something added to the soul to enhance its computing power.[4]

One reason that dualism is intuitive and monism is counterintuitive is that the brain does not perceive its own neural processing, so mental activity is often attributed to some other source—a "mind" or "soul" or "consciousness"—that feels as though it exists separately from the brain. By contrast, most Western-trained scientists, including me, are monists, in the sense that we don't trust our dualistic intuitions, in the same way that we don't trust our intuitions that the earth is stationary and the sun goes around it, even though that is what it feels like and looks like.

Some of the spiritual seekers are not strictly dualist in their approach, with many believing that consciousness, or mind, is primary and everything else is derivative of consciousness or mind. Call that *mind monism*, in contrast with the *matter monism* of most scientists. Deepak Chopra is one such *mind monist*. By contrast, I am a *matter monist*. (There are also Western "idealist" philosophers who hold a type of mind monism, believing that ideas are primary.[5]) In conversations with me, Chopra admitted that the matter monists are far ahead in scientific acceptance, but he counters by noting, "We see mind produce matter every time a synaptic gap is crossed by neurotransmitters or hormones are triggered by experience. A person who is phobic sees a spider and reacts violently at the level of stress hormones, elevated heart rate, elevated blood pressure, etc. This physical state is entirely produced by a mental interpretation that a harmless insect is cause for panic." Deepak also likes to quote famous physicists in support of his mind monism, such as Roger Penrose: "Consciousness is the phenomenon whereby the universe's very existence is known." And Werner Heisenberg: "The atoms or elementary particles themselves are not real; they form a world of potentialities or possibilities rather than one of things or facts."[6] In his 2017 book, *You Are the Universe*, Chopra asserts:

> Consciousness is fundamental and without cause. It is the ground state of existence. As conscious beings, humans cannot experience, measure, or conceive of a reality devoid of consciousness.[7]

Well, yes, that's true by definition. You have to be conscious to experience anything, so when Deepak proposes that consciousness and the universe are equivalent, in the sense that it is an "undeniable fact that any universe is only knowable through the human mind's ability to perceive reality," he is stating the obvious. Call this the *Weak Consciousness Principle* (WCP): *you have to be conscious to experience consciousness*. But Deepak goes further than this when he says that "if all human knowledge is rooted in consciousness, perhaps we are viewing not the real universe based on limitations of the brain" and that "the apparent evolution of the cosmos since the big bang has been totally dependent upon human consciousness." That's reversing the causal arrow, from perception to determination, from being consciously aware of the universe and trying to understand it, to our own consciousness bringing about the universe. Call this the *Strong Consciousness Principle* (SCP). Analogously, if there is no one in the forest observing a falling tree, then the impact on the ground will make no sound, if we define "sound" as the vibration of air stimulating the hearing apparatus of sentient beings. But if we take all conscious beings out of the equation, that doesn't mean trees, atoms, and universes cease to exist. We just have a different definition of sound, trees, atoms, and universes. In like manner, it is one thing to define the existence of atoms or spiders as percepts forming concepts in conscious brains, but that doesn't mean atoms and spiders would not exist without those perceiving brains. Here we are talking at two different levels of analysis, both equally valid but neither one gainsaying the other. They are complementary, not contradictory.

In an attempt to merge Western scientific and Eastern spiritual traditions, Chopra and his colleagues believe they may have found a

pathway to *mind monism* through quantum physics and the neuroscience of consciousness, which he often explores through conferences. At
his 2012 Sages and Scientists symposium, for example, Deepak challenged me to consider that consciousness exists separately from the
brain. I responded with a question: "Where is Aunt Millie's mind when
her brain dies of Alzheimer's?" That is, we know what happens when
the plaques and tangles surround and invade the neurons in an Alzheimer's victim's brain as the disease spreads and kills nerve cells. The brain
shrinks and with it so do the thoughts and memories of Alzheimer's
patients. It's a debilitating disease that serves witness to the necessity of
neurons for thoughts and memories. "Aunt Millie was an impermanent
pattern of behavior of the universe and returned to the potential she
emerged from," Chopra rejoined to my challenge. "In the philosophic
framework of Eastern traditions, ego identity is an illusion and the goal
of enlightenment is to transcend to a more universal, nonlocal, nonma-
terial identity." A few of the quantum physicists at the conference conjectured that consciousness may exist outside the brain in nonlocal
quantum fields, in which subatomic particles appear to be connected
nonphysically in a manner Einstein described as "spooky action at a
distance." But, I suggested aloud, the fact that quantum nonlocality
is spooky and that consciousness is spooky does not mean that they
are causally connected. Spookiness is not a binding substance between
concepts.

Consciousness-first proponents retort that the brain is like a television set and consciousness is like the television broadcast signals. Just
as you need a TV set to receive broadcast signals, you need a brain to
register consciousness. The Dutch near death experience researcher Pim
van Lommel articulates the argument thus:

> We only become aware of these electromagnetic informational fields at
> the moment we switch on our TV, cell-phone, or laptop. What we receive
> is not inside the instrument, nor in the compartments, but thanks to
> the receiver, the information from the electromagnetic fields becomes
> observable to our senses and hence perception occurs in our conscious-
> ness. If we switch off the TV set, the reception disappears, but the
> transmission continues. The information transmitted remains present
> within the electromagnetic fields. The connection has been interrupted,
> but it has not vanished ("non-locality") . . . As soon as the function of

the brain has been lost, as in clinical death . . . memories and consciousness do still exist, but the receptivity is lost, the connection is interrupted.[8]

Thus, says Van Lommel, death does not mean the end of consciousness. The full-throated version of the analogy was made in a 2009 book titled *Irreducible Mind*, edited by Edward and Emily Kelly, in which they argue that "autobiographical, semantic, and procedural (skill) memories sometimes survive bodily death. If this is the case, memory in living persons presumably exists, at least in part, *outside* the brain and body as conventionally understood." How could this be, given what we know about the brain and how memories are stored in neural patterns? "The true function of the brain might for example be *permissive*, like the trigger of a crossbow, or more importantly, *transmissive*, like an optical lens or a prism, or like the keys of a pipe organ (or perhaps, in more contemporary terms, like the receivers in our radios and televisions)."[9]

The analogy doesn't hold. Television studios generate and broadcast signals that our TV sets pick up. If our brains are analogous to television sets, where is the consciousness equivalent to TV production and broadcast facilities? Who or what is doing the consciousness broadcasting? In other words, if brains are not the source of consciousness, then what is? In point of fact there is no consciousness broadcaster and the brain is nothing at all like a television set, and believers in the soul have no answers to these questions beyond a vague notion that consciousness is everywhere. What neuroscience tells us is that everything a mind (or soul) is supposed to do fails when the accompanying part of the brain fails,[10] and that is probably why only 7.1 percent of biologists believe in an afterlife.[11]

THE LANGUAGE PROBLEM

Much of this debate about the nature of consciousness turns on the language of the different worldviews.[12] The words we use to describe the worldview we hold are important to understand in order to clearly communicate, and part of the problem many Western-trained scientists have with Eastern spiritual traditions is the language that, to many of

us, sounds nonsensical. For example, Deepak routinely tweets statements that sound like gobbledygook:

> *In the deeper reality beyond space & time we are members of one body*
> *and one mind.*

> *Consciousness regulates and becomes the flow of energy and informa-*
> *tion in your body.*

If you go to wisdomofchopra.com/quiz.php you can test your ability to discriminate between tweets that are genuine Deepak utterances
and fake messages generated by a computer program. It is often difficult
to tell the difference (e.g., "True identity expresses ephemeral belonging"
is fake). In a 2015 paper by the psychologist Gordon Pennycook and
his colleagues, such tweets are examples of what they called "pseudo-
profound" bullshit, or language "constructed to impress upon the reader
some sense of profundity at the expense of a clear exposition of meaning or truth."[13] I am cited in the paper for describing Chopra's language
as "woo-woo nonsense," which came from a 2010 debate Deepak and I
had at Caltech (with Sam Harris on my side and Jean Houston on
Deepak's side) that was televised on ABC's *Nightline*. In the audience
Q&A, Chopra engaged the physicist and science writer Leonard Mlodinow in a dialogue on the nature of consciousness, which Deepak
defined as "a superposition of possibilities." To this Mlodinow responded:
"I know what each of those words means. I still don't think I know
[what you mean by consciousness]."[14]

Deepak's definition of consciousness here certainly sounds like
pseudo-profound bafflegab, but I have since gotten to know him and
can assure readers that he doesn't create such phrases with intentional
obfuscation. Consciousness remains unexplained to scientists—at least
conscious qualitative experience, or *qualia*—and Deepak and others
believe that quantum physics can help to explain it ("a superposition
of possibilities" describes some subatomic quantum effects). So invoking terms from that field makes sense in his mind, even though to others
much of what he says can sound nonsensical.

If you want people to understand your ideas, you must communicate them clearly, so I have long maintained the position that the burden is on Chopra to convey his ideas clearly. But my wife convinced

me that communication is a two-way street and that to better understand Deepak's words, I needed to enter his world. Hence the journey
to Carlsbad.

INSIDE CONSCIOUSNESS

Our experience at the Chopra Center began with a fairly comprehensive assessment of our personality, lifestyle, diet, and other medically
relevant factors. A consult with the resident medical doctor was followed
by an explanation of Vedic science from our Master Educator, a woman
named Manjula Nadarajah, who explained that the theory behind
the Ayurvedic treatments we were to receive involves three doshas
(types)—Vata, Pitta, and Kapha—and that my dosha is primarily a Pitta:
"medium build, sharp intellect, good decision maker, bright, and
warm." Well, who am I to argue with such insight? Then again, when
out of balance I can be "angry, irritable, and judgmental." Oh.

Balance comes from integrating body, mind, and spirit through diet,
exercise, and meditation. Since Pittas are "hot, sharp, sour, pungent, and
penetrating," to balance that I need to "make choices that are cooling,
sweet, and stabilizing." What does this mean on a practical level? A daily
routine of free time, no unnecessary time pressures, no skipped meals;
favor foods that are sweet, bitter, and astringent, and choose cooling
foods such as cucumbers, sweet fruits, and melons. I am advised to spend
time in nature, take walks in the woods and along natural bodies of
water, get regular massages, and favor aromas that are cooling and
sweet such as sandalwood, rose, jasmine, mint, lavender, fennel, and
chamomile. Oh, and I should laugh many times each day. Well, that
sounds about right for me . . . and for just about everyone else, too.
Aren't these all mostly good things for *anyone* to do? How would anyone *not* feel better following this advice?

One of the principal exercises of spiritual seekers is meditation, said
to be the avenue to deeper consciousness, so I tried three sessions of
meditation with my personalized mantra. As a Pitta, I was told to
employ a "4-7-8" breathing pattern: inhale for four seconds, hold my
breath for seven, then exhale for eight. This I did for half an hour, three
times over the course of the weekend. I'm told it takes many years to
master meditation, so I was merely dabbling here, but for a few minutes,

anyway, I managed to attenuate some of the thought-flooding and negative emotions that cause stress and anxiety.

Best of all were the Ayurvedic massage treatments, one of which,
called Gandharva, included warm oils and a crystal "singing bowl" that
my Healing Arts Master (better known to us Muggles as a massage
therapist) utilized to produce deep sound vibrations that I could feel
pulsating through my body, which, I was told, is explained by Vedic
science. Figure 4-1 shows Manjula with my wife and me in the lobby
of the Chopra Center, along with the Ayurvedic massage oils said to aid
in the adjustment of one's energy, mine being primarily Pitta.

Uncomfortable with the term "Vedic science"—because none of this
is what I think of when I think of science—I pressed Manjula for more
information on how it supposedly works. "During meditation," she
explained, "we expand our internal reference point from local to nonlocal, from constructed to expanded awareness, from a skin-encapsulated
ego to a field of ever-present witnessing awareness." I understand what

Figure 4-1

Chopra Center master educator Manjula Nadarajah with the author and his
wife, along with the Ayurvedic massage oils said to aid in the adjustment of
one's energy. From the author's collection.

the words mean by themselves but am still puzzled by the language. Manjula continued to explain that consciousness is "primary, nontangible, [and] nonlocal" and may be described scientifically as "a quantum mechanical field of interrelatedness." Deepak reiterated the point when I prodded him for further clarification. Brains and minds—and everything else—are different manifestations of conscious awareness: rocks are in quiet awareness, plants are waking up, animals are moving around, and humans are self-aware. Life is consciousness in physical form. Birth and death are transitions of conscious states, entering and departing the physical manifestation of consciousness. The soul is the essence of that particular conscious manifestation. God is consciousness.[15]

The key to understanding this worldview is consciousness. "Vedic science regards consciousness as a fundamental property of the universe," Chopra continued. "Consciousness is where all experience occurs, in which all experience is known and out of which all experience is made." Space, time, energy, information, and matter are all manifestations of consciousness. "Experience is the basis of all the reality we know." In this worldview, then, some of Deepak's seemingly nonsensical tweets begin to make sense. On June 11, 2012, for example, he tweeted, "Consciousness differentiates into space time energy information and matter. Differentiation is not separation. They are one." In a worldview in which mind and consciousness are emergent properties of neurons firing in complex patterns in the brain, consciousness is a secondary property, and so Deepak's tweet makes no sense. But in a worldview in which consciousness is primary, it does.

Which worldview is right: mind monism or matter monism? Wrong question. The scientific paradigm may be correct in describing the physical manifestations of consciousness (rocks, plants, animals), Chopra continued, but "it's incomplete being based in a subject/object split which is artificial." How do we know consciousness is primary? Meditation is one way to find out, Chopra believes, but like introspection it is a solely personal experience and as such is not subject to external validation.

The *effects* of meditation, however, can be measured, and in 2016 Chopra opened his center to scientists from Harvard Medical School, UC San Francisco, and the Icahn School of Medicine at Mount Sinai to run an experiment on the effects of meditation on a number of health measures, including aging biomarkers, stress indicators, and general biological processes, along with self-reports of well-being. At the La Costa Resort and Spa, healthy women aged thirty to sixty were randomly assigned to one of two groups, (1) vacation only (n=31) and (2) novice meditation (n=33), and these were both compared to a third group of regular meditators (n=30) who were already enrolled for the six-day stay at the facility. It is an interesting and important study because it controlled for the vacation effects that anyone would experience at such a posh place, allowing the scientists to compare these to more specific effects of meditation, both novice and experienced.

Predictably, all three groups "felt greater vitality and decreased distress" and showed immediate impact on molecular networks associated with stress and immune pathways. So taking vacations is good for both body and mind. But compared to the vacation group, the novice meditation cohort "exhibited more sustained well-being up to 1 month later," and ten months later "the novices maintained a clinically meaningful improvement in depressive symptoms."

Of course, self-report psychological states are notoriously difficult to interpret because of their subjectivity of measurement and meaning, so the research team also examined changes in twenty thousand genes to determine which changed before and after the resort stay. Both intensive meditation and relaxation vacation led to beneficial changes in gene networks involved with stress and inflammation. For regular meditators, a week of intense meditation led to additional beneficial changes in gene expression and age-related proteins not observed in other groups. Compared to the vacationers, novice meditators had beneficial changes in Alzheimer's-related markers and maintained stress reduction a month later. More specifically, "regular meditators showed post-intervention differences in a gene network characterized by lower regulation of protein synthesis and viral genome activity" and "regular meditators showed a trend toward increased telomerase activity compared with randomized women, who showed increased plasma Aβ42/Aβ40 ratios and tumor necrosis factor alpha (TNF-α) levels."

These latter findings are significant for three reasons: (1) Telomerase is involved in the maintenance of telomeres (more on these later), which enable cells to continue dividing, and shorter telomeres "have been predictive of earlier onset of chronic diseases of aging, including diabetes, cardiovascular disease and certain cancers." (2) Greater Aβ42/Aβ40 ratios are associated with lower risk of dementia and Alzheimer's disease, lower risk of major depression, and greater longevity. (3) TNF-α is involved in the regulation of immune cells in the suppression of tumor cell growth.[16] According to coauthor Rudy Tanzi from Harvard, reflecting on the consequences of this research in an email to me:

> Until now any potential positive effects of meditation or a relaxing vacation have been considered to be purely psychological with regard to stress reduction. We have shown that these benefits have a physical origin involving changes in programs of gene expression as well as biochemical events that are predicted to be beneficial. In essence, genes that are normally in high gear to protect you, e.g. inflammation- and infection-related ones, begin to stand down. Since excess activity of these genes can lead to tissue damage and deterioration, dampening down of these genes owing to regular meditation should lead to a healthier condition, in addition to the psychological and spiritual benefits, which also appear to be long-lasting.[17]

If these effects are replicated in future studies, it will strengthen the mind-body connection these practitioners believe is a powerful one, but we should keep in mind that people who meditate regularly probably engage in many other lifestyle activities that lead to such salubrious effects. Perhaps in day-to-day life, regular meditators also monitor their diet more judiciously, smoke and drink less, exercise more, and are risk averse.[18] Still, as Deepak reassured me, "You don't have to buy into the philosophy to benefit." That's good, because I did benefit, even while remaining skeptical that consciousness is the ground of being and the fundamental constituent of the universe.

LIFE AND AFTERLIFE

What does this have to do with death, immortality, and heaven? According to Chopra, the answer has to do with qualia, that is, the qualitative experiences of life. "All subjective experiences are qualia. The experience of the body is a qualia experience. The experience of mental activity is a qualia experience." This qualia is there before birth and continues after death. "Birth is the beginning of a particular qualia program. Death is the termination of a particular qualia program. The qualia return to a state of potential forms within consciousness, where they reshuffle and recycle as new living entities. The consciousness field and its matrix of qualia are nonlocal and immortal."[19] Thus, a proper understanding of mortality and immortality depends on comprehending consciousness. As Deepak writes in his book *Life After Death*, "since by definition death brings physical life to an end," in order to fully grasp the proof of life after death "we must expand the boundaries of consciousness so that we know ourselves better. If you know yourself as someone beyond time and space, your identity will have expanded to include death."[20]

This does not follow from what we understand from neuroscience, which is that our minds, like our souls, are in our brains. Damage to the fusiform gyrus of the temporal lobe, for example, causes face blindness, and stimulation of this same area causes people to see faces spontaneously. Stroke-caused damage to the visual cortex region called V1 leads to loss of conscious visual perception. Patients suffering from strokes in other parts of their brain have lost the capacity for emotions and even for decision making. Damage to the prefrontal cortex results in high risk-taking and rule-breaking behavior. A famous case of a man who suddenly developed pedophilic feelings was discovered to have a tumor at the base of his orbitofrontal cortex that pressed up against the right prefrontal region of his brain, an area associated with impulse control. When the tumor was resected, he lost all pedophilic feelings. When they returned months later, it was discovered that the tumor had grown back.

Changes in conscious experience can be directly measured by fMRI, EEG, and single-neuron recordings. Neuroscientists can predict human choices from brain scan activity before the subject is even consciously aware of the decisions made. Using brain scans alone, neuroscientists have even been able to reconstruct on a computer screen what someone is seeing. *Brain activity=conscious experience*. Thousands of lab experiments, in conjunction with naturally occurring experiments in the form of brain tumors, strokes, accidents, and injuries, confirm the hypothesis that neurochemical processes produce subjective experiences. *Neural activity=qualia*. The fact that neuroscientists are not in agreement over which physicalist theory best accounts for the mind does not mean that the hypothesis that consciousness creates matter holds equal standing.

No one denies that consciousness is a knotty problem. But before we reify consciousness to the level of independent agency capable of creating its own reality separate from the brain, let's give the hypotheses we do have for how brains create minds more time. We know for a fact that measurable consciousness dies when the brain dies, so until proven otherwise, the default hypothesis should be that brains cause consciousness. *I am, therefore I think.*

———

SPIRITUAL SEEKERS AND believers in the soul respond that, in fact, scientific evidence does exist for the afterlife in the form of near death experiences and reincarnation, which we will examine in detail in the next chapter. As I will demonstrate, as intriguing as these stories are, they do not amount to adequate evidence of an afterlife.

CHAPTER 5

EVIDENCE FOR THE AFTERLIFE

Near Death Experiences and Reincarnation

To be mortal is the most basic human experience and yet man has never been able to accept it, grasp it, and behave accordingly. Man doesn't know how to be mortal.

—Milan Kundera, *Immortality*, 1990[1]

Most religious instantiations of the afterlife, such as the visions of heaven proffered by Judaism, Christianity, and Islam, are articles of faith to be accepted without demand for evidence or proof. The scientific quest for immortality, however, is predicated on the belief that evidence is not only central but in fact already exists in the form of near death experiences and reincarnation. Let's examine both of these stairways to heaven independently, as each offers a different explanation for what is really going on.

NEAR DEATH EXPERIENCES AS STAIRWAYS TO HEAVEN

Near death experiences (NDEs) are typically characterized by five common components: (1) an out-of-body experience (OBE) with the feeling of floating above one's body and looking down; (2) separation from the body; (3) entering darkness through a tunnel or hallway; (4) seeing a bright light at the end of the tunnel that serves as a passageway to . . . ;

(5) the "other side," where light, God, angels, loved ones, and others who have "passed over" are there to welcome the dying person.

Sometimes there is a life review, and though most NDEs are positive and lead people to experience gratitude and joy, according to the International Association of Near-Death Studies, 9 to 23 percent of people have had *negative* NDEs characterized by fear, void, emptiness, pain, and even nonexistence. Instead of going to heaven, some of these people find themselves in hell.[2] According to an NDE researcher named Phyllis Atwater, who has had NDEs herself and specializes in the negative experiences some people report, the hellish NDEs are experienced by "those who seem to have deeply repressed guilt, fear, and anger, or those who expect some kind of punishment after death."[3] In other words, when we attempt to explain NDEs we must recognize that there is a wide variety of them and so no one monolithic theory can account for them all, whatever they actually represent.

NDEs and OBEs arose in public consciousness in 1975 through Raymond Moody's bestselling book *Life After Life*, which recounted over a hundred such cases, which many people took to be evidence of an afterlife. The rate or frequency of NDEs is difficult to pin down with reliable numbers. A cardiologist named Fred Schoonmaker, for example, reported that 50 percent of more than two thousand of his patients over an eighteen-year period reported NDEs.[4] A 1982 Gallup poll, however, reported a percentage an order of magnitude smaller at 5 percent.[5] Another cardiologist, Pim van Lommel, claims that 12 percent of his 344 cardiac arrest patients who were successfully revived had NDEs,[6] and in his book *Consciousness Beyond Life* he echoes what most people believe—NDEs are evidence of the survival of the mind without a brain.[7]

The most famous NDE occurred in 1984 when a migrant worker named Maria was hospitalized in Seattle after a heart attack. There in the ICU she suffered another cardiac arrest. After being resuscitated she reported that she floated out of her body up to the ceiling from where she could observe medical personnel working on her. Most remarkably, she says she then journeyed outside the hospital room, where she saw a tennis shoe on the ledge of a third-floor window. Her ICU social worker, a woman named Kimberly Clark, says she went up to the third floor and found a shoe on a window ledge: "The only way she could have had such a perspective was if she had been floating right outside and at very close range to the tennis shoe. I retrieved the shoe and brought it back to Maria; it was very concrete evidence to me."[8] Evidence of what, exactly? A slew of bestselling books in recent years lays out exactly what these experiencers believe NDEs are proof of and where they went during their trip: *Heaven Is for Real*, *To Heaven and Back*, *The Boy Who Came Back from Heaven*, and most notably *Proof of Heaven: A Neurosurgeon's Journey into the Afterlife* by the Harvard neurosurgeon Eben Alexander.

HUME'S MAXIM APPLIED TO NDES

Proof. That's a strong word. Do NDEs represent proof of an afterlife? We can frame this question as the great Scottish Enlightenment philosopher David Hume did in his analysis of miracles in his 1758 work *An Enquiry Concerning Human Understanding*. In it Hume introduces a maxim to apply whenever one comes across an account of an apparently supernatural occurrence, such as a miracle:

> The plain consequence is (and it is a general maxim worthy of our attention), "That no testimony is sufficient to establish a miracle, unless the testimony be of such a kind, that its falsehood would be more miraculous than the fact which it endeavors to establish."

What's more likely? Miracles, or that people's *accounts* of miracles are mistaken? We have very little evidence for miracles, but we have lots of evidence that people misunderstand, misperceive, exaggerate, or even make up stories about what they think they witnessed or experienced. Hume's example of a miracle is the resurrection of the dead. What's more likely—that dead people can come back to life, or that the accounts of dead people being resurrected are in error? Hume answers the question this way:

> When anyone tells me that he saw a dead man restored to life, I immediately consider with myself whether it be more probable, that this person should either deceive or be deceived, or that the fact, which he relates, should really have happened. I weigh the one miracle against

the other; and according to the superiority, which I discover, I pronounce my decision, and always reject the greater miracle. If the falsehood of his testimony would be more miraculous than the event which he relates; then, and not till then, can he pretend to command my belief or opinion.[9]

Applying Hume's Maxim to NDEs, we can inquire, which is more miraculous: the falsehood of the accounts of NDEs or what they purportedly represent? And we can ask ourselves what's more likely: that NDE accounts represent descriptions of actual journeys to the afterlife or are portrayals of experiences produced by brain activity? Many lines of evidence converge to support the theory that NDEs are produced by the brain and are not stairways to heaven. Let's look at these lines of evidence in detail, starting with Hume's recognition that people can either deceive or be deceived.

NDES AS FICTIONAL ACCOUNTS

I have an adage that I impart to my students and audiences: *Sometimes people just make things up.* It's called fiction. *The Lord of the Rings, The Chronicles of Narnia,* the *Harry Potter* series, the *Star Wars* saga. Fake, fake, fake, fake. They're all fiction and no one confuses them with nonfiction. That would be like mistaking Dante Alighieri's 1320 poem *The Divine Comedy,* considered one of the preeminent works of Western literature in its imaginative vision of the afterlife, for an actual account by someone who went there and returned to report on what he saw. The illustration by Gustave Doré (figure 5-1) for an 1892 edition of *The Divine Comedy* attempts to portray the empyrean of God as envisioned by medieval Christian theologians, which inspired the work, but no one thinks that this is an actual depiction of heaven.

The fact is that humans have a remarkable capacity to create the most fantastic tales in vivid detail that go on for paragraphs, pages, chapters, books, and series. Adding graphic details to a story does not elevate it to verisimilitude. NDE accounts that include specifics like the tennis shoe on the window ledge, or that the other side beyond the bright light at the end of the tunnel is rich in vibrant colors, sonorous sounds, or effervescent environments, are to my ears no different from

Figure 5-1. The Empyrean of God

Dante Alighieri's 1320 poem *The Divine Comedy* is an imaginative vision of the afterlife inspired by medieval Christian theologians. The artist Gustave Doré illustrated God's empyrean for an 1892 edition of the work.

the elaborate stories I hear from people who claim they were abducted by aliens and recount the particulars of the spaceship interior. So what? George Lucas's imagination worked wonders for the inside of Han Solo's *Millennium Falcon* or the Empire's Death Star. Adding "nonfiction" to a book title doesn't make it true. In this light, it is revealing that the author of *The Boy Who Came Back from Heaven*, improbably named Alex Malarkey, recanted his allegedly true story, admitting that he made it all up.[10]

What about that remarkable story of the tennis shoe sighting on the window ledge during an NDE? First, we have only the word of Maria and her social worker that it happened at all, and as the investigative journalist Gideon Lichfield noted when he tried to chase down the story for an article in the *Atlantic* on "The Science of Near-Death Experiences," the story was "thin on the evidential side," and he learned when he tried to verify the account that "A few years after being treated, Maria disappeared, and nobody was able to track her down to further confirm her story."[11]

Because the tennis shoe tale has taken on iconic status in NDE circles, there is now an ongoing experiment by Sam Parnia and others in rooms located in fifteen different hospitals in the United States, the United Kingdom, and Austria where cardiac arrest patients will likely undergo resuscitation efforts. They placed images high on a shelf facing the ceiling so that if an OBE occurs during an NDE and the patient is "up" by the ceiling looking down, they should be able to see the image and later report what it is. So far the researchers have recorded a total of 2,060 cardiac arrests, 330 survivors, 140 interviewees, 9 remembered NDEs, and only 1 OBE, the patient in this case saying he floated up to a corner of the room where he watched the staff try to resuscitate him. Parnia and his colleagues concluded that the details of his description were eerily accurate. Eerily accurate as to what? Not the description of the image on a shelf, because there was none in this particular room at the time. In fact, the description was of the doctors working on him. But most of us have seen television shows and movies depicting doctors using a defibrillator to start someone's heart after cardiac arrest, so any description even remotely close is going to sound "eerie" to those who want NDEs to be real.

NEAR DEATH IS NOT DEAD

Any explanation for the NDE must begin with the fact that there's a reason why the N modifies the D: the people who experience them are *not actually dead*. They're only *near* death, a state in which the brain may undergo stress, be deprived of oxygen, release neurochemicals that can mimic the hallucinatory trips of drug users, or experience any one of the dozens of anomalous neurological anomalies, abnormalities, or disorders that have been documented by neurologists and neuroscientists. The fact that each NDE is unique does not mean that some of them are real trips to heaven (or hell) while the rest are merely by-products of a hallucinating brain. It just means that the brain is capable of a wide variety of experiences depending on the immediate conditions and one's personal life trajectory, all of which are necessarily unique but no less caused by internal brain states.

In their NDE accounts, experiencers will often emphasize that they were "dead" or "absolutely dead" or "clinically dead" in order to bias the interpretation toward the miraculous or supernatural. A Portland, Oregon, emergency room doctor named Mark Crislip, however, reviewed the original EEG readings of a number of patients claimed by scientists as being flatlined or dead and discovered that they weren't dead at all. "What they showed was slowing, attenuation, and other changes, but only a minority of patients had a flat line, and it took longer than 10 seconds. The curious thing was that even a little blood flow in some patients was enough to keep EEGs normal." Crislip also analyzed the NDE study by Pim van Lommel and his colleagues published in the prestigious British medical journal *Lancet*, in which the authors "defined clinical death as a period of unconsciousness caused by insufficient blood supply to the brain because of inadequate blood circulation, breathing, or both. If, in this situation, CPR is not started within 5–10 min, irreparable damage is done to the brain and the patient will die."[12] As Crislip notes, however, most of these cardiac patients *were* given CPR, which by definition delivers oxygenated blood to the brain (that's the whole point of doing it). "By the definitions presented in the *Lancet* paper, nobody experienced clinical death," Dr. Crislip concluded, adding that as a physician who has conducted CPR many times, "No doctor would ever declare a patient in the middle of a code 99 dead, much less brain dead. Having your heart stop for 2 to 10 minutes and being promptly resuscitated doesn't make you 'clinically dead.' It only means your heart isn't beating and you may not be conscious."[13]

So proponents' claim that in NDEs people die and then travel to the other side is gainsaid by the fact that they never actually died.

NDES AS HALLUCINATIONS

To my ears many of these NDE accounts are indistinguishable from those of people who have had drug-induced hallucinatory trips. Take Eben Alexander's NDE story of his "trip" to the afterlife during a meningitis-induced coma.[14] There he met a young woman with high cheekbones, deep-blue eyes, and "golden brown tresses" framing her face. Together they traveled on the wing of a butterfly; "in fact, millions of butterflies were all around us—vast fluttering waves of them, dipping down into the woods and coming back up around us again. It was a river of life and color, moving through the air." The woman's outfit "was simple, like a peasant's, but its colors—powder blue, indigo, and pastel orange-peach—had the same overwhelming, super-vivid aliveness that everything else had." Alexander was then overwhelmed with a feeling of love, not friendship or romantic but "somehow beyond all these, beyond all the different compartments of love we have down here on earth. It was something higher, holding all those other kinds of love within itself while at the same time being much bigger than all of them." Her message to him was simple: "You are loved and cherished, dearly, forever."

Compare Eben Alexander's trip with the "trip" taken by the neuro-scientist Sam Harris after he and a friend ingested a dose of the drug MDMA, better known as Ecstasy, which he details in the opening pages of his book *Waking Up*.[15] Harris reports that he was "suddenly struck by the knowledge that I loved my friend." Not friendship or roman-tic, but "this feeling had ethical implications that suddenly seemed as profound as they now sound pedestrian on the page: *I wanted him to be happy*." More than this, Harris says, "came the insight that irrevoca-bly transformed my sense of how good human life could be. I was feel-ing boundless love for one of my best friends, and I suddenly realized that if a stranger had walked through the door at that moment, he or she would have been fully included in this love."

The "love" theme appears to be a common one among NDEs, as well as other anomalous psychological experiences, such as the one I wrote about in *The Believing Brain* that happened to my friend Chick D'Arpino at four in the morning on February 11, 1966. When he was alone in a bedroom at his sister's home, feeling despair and loneliness while going through a painful divorce involving the custodial loss of his children, all of a sudden he heard a voice that was neither masculine nor feminine and seemed to him to come from out of this world. It was so powerful a message that Chick took it upon himself to deliver it to President Lyndon Johnson at the White House, a journey that landed him in a mental institution instead. Although Chick has never told anyone the precise words of the message, or what he thinks the source was, its essence, he told me, was *love*. "The source not only knows we're here, but it loves us and we can have a relationship with it."[16]

Psychedelic drugs can have similar emotional effects. In recent years psychedelics have been experimented with for the treatment of depression, PTSD, and death anxiety in terminally ill patients. A single dose of psilocybin (LSD) given to cancer patients at NYU Medical School by the psychiatrist Stephen Ross, for example, not only reduced their depression and anxiety, but the effects were so dramatic that Ross said, "I thought the first ten or twenty people were plants—that they must be faking it. They were saying things like 'I understand love is the most powerful force on the planet' . . . The fact that a drug given once can have such an effect for so long is an unprecedented finding. We have never had anything like it in the psychiatric field."[17]

Another instructive comparison with Eben Alexander's NDE are the drug-induced trips the late neurologist Oliver Sacks took in his life, recounted in his autobiography, *On the Move*. In November 1965, for example, Dr. Sacks was putting in marathon workweeks and downing huge doses of amphetamines to stay awake, topped off with generous measures of sleep-inducing chloral hydrate. One day while dining in a café, as he was stirring his coffee, "it suddenly turned green, then purple." When Sacks looked up he noticed that the customer at the cash register "had a huge proboscidean head, like an elephant seal." Shaken by this image, Sacks ran out of the diner and across the street to a bus, where all the passengers "seemed to have smooth white heads like giant eggs, with huge glittering eyes like the faceted compound eyes of insects." At that moment the neurologist realized that he was hallucinating but that "I could not stop what was happening in my brain, and that I had to maintain at least an external control and not panic or scream or become catatonic, faced by the bug-eyed monsters around me."[18]

I have appeared on a number of television shows with Eben Alexander and have spent a fair amount of time with him in green rooms before and after, discussing what he thinks happened to him. I enjoyed our conversations and found him to be an amiable man, but he is a neurosurgeon who knows the literature about hallucinations and the many tricks the mind can play under a wide variety of conditions. Why doesn't Alexander recognize his own experience as something similar to what so many others have undergone who don't claim to have tripped off to celestial empyreans? Because it happened to *him*, and subjective experiences are orders of magnitude more powerful than anything you can read in a book.

There are additional problems with Alexander's claims. During his NDE he says that his "cortex was completely shut down." He concludes from this that "there is absolutely no way that I could have experienced even a dim and limited consciousness during my time in the coma," and therefore "my brain-free consciousness journeyed to another, larger dimension of the universe." According to Dr. Laura Potter, the attending physician the night he was wheeled into the ER, however, Alexander's coma was induced by her in order to keep him alive while he was heavily medicated, and that whenever they tried to wake him he thrashed about, pulling at his tubes and trying to scream, so his brain was not completely shut down. When Potter later challenged him on this point, Alexander told her his account was "artistic license" and "dramatized, so it may not be exactly how it went, but it's supposed to be interesting for readers."[19] In other words, Alexander mashed fact and fiction, meaning that there is really nothing to be explained.

We now know of a number of factors that produce such fantastical hallucinations, also masterfully explicated by Oliver Sacks in his book *Hallucinations*. For example, Sacks recounts an experiment by the Swiss neuroscientist Olaf Blanke and his colleagues, who produced a "shadow-person" in a patient by electrically stimulating the left temporoparietal junction in her brain. "When the woman was lying down a mild stimulation of this area gave her the impression that someone was behind her; a stronger stimulation allowed her to define the 'someone' as young but of indeterminate sex." Sacks also recounts his experience treating eighty deeply Parkinsonian postencephalitic patients (as featured in the film *Awakenings*, starring Robin Williams as Sacks), noting, "I found that perhaps a third of them had experienced visual hallucinations for years before L-dopa was introduced—hallucinations of a predominantly benign and sociable sort," speculating that "it might be related to their isolation and social deprivation, their longing for the

world—an attempt to provide a virtual reality, a hallucinatory substitute for the real world which had been taken from them."[20]

Migraine headaches also produce hallucinations, which Sacks himself experienced as a longtime sufferer, including a "shimmering light"
that was "dazzlingly bright": "It expanded, becoming an enormous
arc stretching from the ground to the sky, with sharp, glittering, zig-
zagging borders and brilliant blue and orange colors." Compare
Sacks's experience to Eben Alexander's trip to heaven during which he
says he was "in a place of clouds. Big, puffy, pink-white ones that
showed up sharply against the deep blue-black sky. Higher than the
clouds—immeasurably higher—flocks of transparent, shimmering
beings arced across the sky, leaving long, streamerlike lines behind
them." The similarities are as unmistakable as they are explicable by
neurochemical changes in their respective brains. There may even be
an evolutionary basis for the sensed presence of others, as Sacks con-
jectures: "Thus the primal, animal sense of 'the other,' which may
have evolved for the detection of threat, can take on a lofty, even
transcendent function in human beings, as a biological basis for religious passion and conviction, where the 'other,' the 'presence,'
becomes the person of God."

In a December 2012 article in the *Atlantic* analyzing Alexander's
claims, Sacks explained that the reason hallucinations seem so real "is
that they deploy the very same systems in the brain that actual perceptions do. When one hallucinates voices, the auditory pathways are activated; when one hallucinates a face, the fusiform face area, normally
used to perceive and identify faces in the environment, is stimulated."
From these facts the neurologist concluded: "The one most plausible
hypothesis in Dr. Alexander's case, then, is that his NDE occurred not
during his coma, but as he was surfacing from the coma and his cortex was returning to full function. It is curious that he does not allow
this obvious and natural explanation, but instead insists on a supernatural one."[21]

Here again we face Hume's question of what's more likely—that
Alexander's NDE was a real trip to heaven and all these other hallucinations are the product of neural activity only, or that *all* such experiences
are mediated by the brain, but to each experiencer they seem real? To me,
this is proof of hallucination, not heaven.

NDES AS BRAIN ANOMALIES

Beyond hallucinations, there are other conditions that may trigger an NDE. The psychologist Susan Blackmore, for example, notes that the "tunnel" effect of the NDE and OBE may be the result of stimulating the visual cortex on the back of the brain where information from the retina is processed. Hypoxia (oxygen deprivation) may interfere with the normal rate of firing by nerve cells in the visual cortex, which may be interpreted by other areas of the brain as concentric rings or spirals, which may be described as a tunnel.[22] In like manner, in an aptly titled book *The Spiritual Doorway in the Brain,* the neurologist Kevin Nelson contends that the tunnel effect may be caused by compromised blood pressure in the eyes during stress or trauma that triggers constriction of the visual fields, and overstimulation of the visual excitation pathway from the brainstem to the visual cortex (the pons-geniculate-occipital or PGO pathway) leads to the bright light sensation.[23]

Another part of the brain implicated in NDEs and OBEs is the right angular gyrus in the temporal lobe, located above and behind the ears. During surgery on a forty-three-year-old woman suffering from epileptic seizures, Olaf Blanke and his colleagues electrically stimulated this neural module and found that when they did so, the woman, now awakened, reported that she could "see myself lying in bed, from above, but I only see my legs and lower trunk." Stimulating an adjacent point in this area induced "an instantaneous feeling of 'lightness' and 'floating' about two meters above the bed, close to the ceiling." The scientists found that through the level of electrical stimulation they could even control the height that the patient reported feeling above the bed. Touching different points in the right angular gyrus produced the sensation that her legs were "becoming shorter" or "moving quickly towards her face," causing her to take "evasive action." The neuroscientist team concluded: "These observations indicate that OBEs and complex somatosensory illusions can be artificially induced by electrical stimulation of the cortex," and that "It is possible that the experience of dissociation of self from the body is a result of failure to integrate complex somatosensory and vestibular information."[24]

An Air Force physician named Dr. James Whinnery discovered what he called "G-Force Induced Loss of Consciousness" (G-LOC) when he accelerated pilots in a training centrifuge to the point of their blacking out from oxygen deprivation. During the fuzzy boundary between consciousness and unconsciousness, many of these pilots experienced brief episodes of tunnel vision, sometimes with a bright light at the end of the tunnel, as well as a sense of floating, sometimes paralysis, and often euphoria and a feeling of peace and serenity when they came back to consciousness.[25] When Dr. Whinnery subjected his charges to G-LOC in a stepwise gradual fashion by slowly accelerating the centrifuge, they experienced tunnel vision, then blindness, then they blacked out because of the loss of oxygen first to the retina, then to the visual cortex, then to the rest of the brain.[26]

As for the "accurate" descriptions of what went on in an OR while the patient was "under," the phenomenon of "anesthesia awareness" happens to about one in a thousand patients, in which they are not totally unconscious during surgical anesthesia. In such cases they may be vaguely aware of what is happening around them, and if they are in a teaching hospital the physicians or surgeons may be narrating the surgery for residents in attendance, which could enable the patient to give a semi-accurate description of events that, when recounted in books and articles later, sounds like watching the procedures from on high.

Finally, there is the problem of incongruence in NDE accounts. Much is made of their similarities—even to the point of being identified as the "invariance hypothesis"—but in point of fact NDE narratives vary considerably, especially across cultures, most notably between Western and Eastern traditions. In India, for example, NDEs rarely include OBEs, or the sensation of passing through a tunnel, or a life review, or the desire to return to the land of the living. In the words of Cory Markum, who documented the problem of incongruity in NDEs: "There is a problem here for people that want to use NDEs as evidence of an objective afterlife: these accounts do not actually seem to be describing the same place." In fact, Markum notes, NDEs would be suggestive of a real heaven "if say, Muslims, atheists, Hindus, and so on *all* returned from a distinctly *Christian* heaven, speaking of Jesus Christ and the Holy Trinity." But they don't. "Instead, what we seem to have is exactly what we would expect if NDEs were the product of the inner workings of the brain. Christians see the figure of Jesus, Hindus see Yamraj and his minions, children's NDEs often seem much more simplistic than those of adults, etc."[27]

REINCARNATION AS A STAIRWAY TO HEAVEN

A second set of evidences often presented as proof of immortality and the afterlife, emanating primarily from the Eastern traditions of Buddhism and Hinduism, is reincarnation, from the ancient Indian Sanskrit word *sansara* for "wandering" or "cyclicality," or the Latin for "entering the flesh again," or in more modern incarnations as the "transmigration of souls." *Dualist* reincarnation holds that upon death the soul departs the body and transmigrates to another body, whereas *mind monism* reincarnation contends that the soul simply returns to the cosmic consciousness whence it came.

The major monotheisms of Judaism, Christianity, and Islam, while believing in souls that migrate from their earthly bodies to a heavenly (or hellish) afterlife, mostly reject the doctrine of reincarnation, although Hollywood has embraced it in such films as *The Search for Bridey Murphy, The Reincarnation of Peter Proud*, and *Audrey Rose*. The latter stars Anthony Hopkins, whose character comes to believe that a girl named Ivy Templeton is the reincarnation of his daughter Audrey Rose, who died in a fiery car crash moments before Ivy was born. In the end Ivy, too, dies while reliving the accident through hypnotic guided imagery, and the ashes of both girls are sent to India for burial as a quote from the Bhagavad Gita scrolls across the screen's final scene:

> *There is no end. For the soul there is never birth nor death. Nor, having once been, does it ever cease to be. It is unborn, eternal, ever-existing, undying and primeval.*

The Hindu Bhagavad Gita, compiled around 500 B.C.E., lays out the doctrine of reincarnation in the context of a great battle in which many lives were lost. As Krishna tells Arjuna:

> If any man thinks he slays, and if another thinks he is slain, neither knows the ways of truth. The Eternal in man cannot kill: the Eternal in man cannot die. He is never born, and he never dies. He is in Eternity: he is for evermore. Never-born and eternal, beyond times gone or to come, he does not die when the body dies. When a man knows him as never-born, everlasting, never-changing, beyond all destruction, how can that man kill a man, or cause another to kill?[28]

The doctrine of reincarnation, at least as it is presented in the Bhagavad Gita, is an understandable response to the tragedies of loss, death, and grief that come with war. If your compatriots are not really dead it perhaps attenuates grief.

Although different religious traditions vary in details about what, exactly, is reincarnated, when, where, and why, the general idea is a cycle of time and an eternal return that involves an ethical/justice component of karma or a "karmic cycle" based on cumulative virtues and vices. In this sense reincarnation is a type of cosmic justice in which the scales are ultimately balanced, or life redemption in which wrongs are righted and the crooked is made straight, and it fits squarely into the *Law of Karma*, which holds that the world is *just* so *justice* will prevail sooner (in this life) or later (in the next life). Everything happens for a reason. There is no cause to grieve over the death of a loved one, for such lives are only temporary stages in the vast drama of the life to come. As Krishna continues: "The wise grieve not for those who live; and they grieve not for those who die—for life and death shall pass away."[29]

A first difficulty with reincarnation is what might be called the *geography problem*. If reincarnation is real, it means that souls in search of new bodies are migrating primarily in and around the Indian subcontinent. This alone should be a red flag for any discerning observer, a strong indication that such beliefs are culturally determined and have no basis in reality. It would be like traveling to India and discovering that physics is entirely different there. There is no "Indian physics" that differs from "British physics." There is just physics because the theories of the science correspond to the facts about the world it studies—a type of correspondence theory of truth that doesn't exist for religious doctrines like reincarnation.

A second obvious impediment with reincarnation is the *population problem*, as evidenced in the population figures with which I began this book. The ratio of the dead to the living is about 14.4 to 1; of the approximately 108 billion people ever to have been born, only about 7.5 billion of them live today. Assuming the 7.5 billion living bodies contain souls from previous people, where are the other 100.5 billion souls? And if the 7.5 billion people today were born with souls—as surely they would be, since that's the theory about people from the past—what happened to their original souls? Were they kicked out of their bodies and left to wander about until they find an open body, or do the living have, on average, 14.4 souls inside them from the previous living?

A third objection involves *personal identity*. If you—your self, your pattern of information representing your thoughts and memories—are carried by the soul and survives death, then what need is there for a body in the first place? If bodies are needed, then each physical incarnation of a soul should be as unique and special as the soul itself, which would obviate reincarnation as a viable phenomenon because wandering or migrating souls mean that bodies are nothing more than temporary vessels as dispensable as clothing.

Beyond the religious, theological, and philosophical arguments there are those who insist that there is *empirical evidence* for reincarnation. In his book *Past Lives, Future Lives* the hypnotherapist and past lives regressionist Bruce Goldberg claims that through hypnosis he can communicate with these lost souls.[30] He is also a future life progressionist, and through hypnosis he says he met the individual who discovered time travel in the year 3050, a man named Taatos. If you haven't noticed any time travelers around it is because "they have mastered hyperspace travel between dimensions and can move through walls and solid objects. By existing in the fifth dimension, they can observe us and remain invisible."[31]

There are numerous problems with this theory. First, hypnosis is not a reliable guide to memory recovery. As the psychologist Elizabeth Loftus has shown time and again in experimental settings and real world cases, people's memories can be easily manipulated by simple suggestion. In testimony regarding an automobile accident, for example, the choice of adjectives used to describe it—say, "smashed" instead of "collided"—influenced the witnesses' estimates of the speed at which they remembered the cars traveling.[32] Loftus's most famous experiment involved planting a false memory in an adult of getting lost in a mall as a child. A third of her subjects "remembered" being lost in the mall, most filling in rich details of what the mall looked like, what happened and when, and even the emotions of being lost and then found.[33] In a similar series of experiments with past life regression, the psychologist Nicholas Spanos demonstrated that there was a strong correlation between believing in reincarnation and the richness and detail of the past life memories, and that the hypnotized subjects were not recalling memories at all but constructing fantasies "as if" they were in a past life, as evidenced by the fact that the people under hypnosis drew upon the suggestions of the hypnotist along with images and information from films, television shows, and novels about reincarnation.[34]

A second line of claimed empirical evidence for reincarnation comes from the research by the late University of Virginia psychiatrist Ian Stevenson, whose massive 2,268-page two-volume 1997 work, *Reincarnation and Biology*, purports to document eerie parallels between living and dead persons, most notably through birthmarks, birth defects, scars, memories, déjà vu experiences, and heaps of anecdotes recounted by those who believe in reincarnation or who are not particularly discerning.[35] Such stories are legion among reincarnation believers, but one need not read deep into the literature to see this process as a classic case of *patternicity*—the tendency to find meaningful patterns in both meaningful and random noise. Real randomness, in fact, is difficult for most of us to grasp intuitively because it results in clusters that look like patterns to the untrained eye. Toss a fistful of coins into the air and note their final resting place on the ground—they won't be perfectly distributed with even spaces between them. They will be congregated here and there in clusters. Ask people to imagine a sequence of coin tosses and they will suggest outcomes like HTHTHTHTHTHTHT, whereas real coin tosses result in streaks of heads and tails, like HHTTTHHHTTTTTH. Early adopters of Apple's "random" shuffle option for its iPod music playlist complained that certain songs repeated more than their intuition allowed, leading the computer company to modify the program to be less random.

My reading of the many reincarnation case studies compiled by Ian Stevenson and others in which birthmarks, birth defects, scars, memories, déjà vu experiences, and the like are believed to be nonrandom and meaningful, is that they are (1) misperceiving the natural clustering of randomness as more significant than it is, (2) finding specific patterns where none exist (A appears to be connected to B but it's not—the false positive or Type I error, which we will discuss in more detail later), and (3) failing to establish an agreed-upon protocol for determining what constitutes significance in such apparent connections. This final point is a methodological problem that has long plagued paranormal research. Whenever possible, psychologists *operationally define* the topic of study so that it can be properly measured and subjected to statistical analysis. A psychic who says "I sense the presence of a father figure here" may appear to get a hit if the subject lost a father, grandfather, uncle, family friend, or anyone who was "fatherly" in nature, but neither "father figures" nor "presences" are measurable.

In reincarnation research, for example, a child's birthmark, birth defect, or scar is "connected" to a fatal injury of a long-dead soldier in that particular body spot. Ian Stevenson, for example, has even computed the odds of birthmarks appearing in one area of a child's body as matched to the wounds of a dead soldier. But how many such marks constitute a hit—one, two, ten? And how close do they need to be to count as a hit? Millimeters? Centimeters? For one set of cases, Stevenson claims that he used a 10-centimeter square as his criterion for a match between wound site and birthmark, but in a careful analysis of Stevenson's data, the philosopher Leonard Angel found it almost impossible to check whether this criterion was actually applied, and where he could check, Angel found that "Stevenson's basic 10 cm sq. match criterion is quite simply not met, with descriptions provided such as 'lower abdomen,' which is far too general." Stevenson also tries to match children's birthmarks to any number of past events: (1) a site of trauma or wound on a deceased person; (2) a site of surgery; (3) a prominent scar; (4) an "experimental mark" (a smudge of ash or soot deliberately made on a corpse prior to interment); (5) a prominent birthmark or skin mark on the past life; (6) a bodily defect, e.g., missing fingers or toes or a congenital defect; (7) the location of an animal bite; (8) a tattoo; (9) a bullet lodged inside the body. As Angel notes:

> By failing to consider the multiplicity of such sites on a typical deceased person Stevenson has fallen into an elementary statistical fallacy similar to the "birthday fallacy." (People untutored in probabilities underestimate the likelihood that two people in a group of 35 will have the same birthday.) Stevenson notices, say, that a birthmark corresponds to a wound mark, but fails to recognize that if the birthmark had been elsewhere it might have corresponded to some other scar, mark, birthmark, birth defect, tattoo, or experimental mark. This is the classic set-up for a statistical blunder and Stevenson has fallen right into it.[36]

And then there is the matter of the unexamined cases and the law of large numbers. Stevenson and other reincarnation researchers usually begin their examination of a case *after* the child's family has decided who they think the reincarnated dead person is—say, a soldier who was shot in the chest or head in a spot that appears to correspond to a scar or birthmark on the child's chest or head. But any child with a birthmark, scar, or birth defect on their chest or head (or anywhere, really) will inevitably appear to replicate the point of fatal impact of some dead soldier at some point in the past if you examine a large enough sample. If you rummage around through enough past people you're bound to find one that appears to fit the pattern. In fact, it would be surprising if there were *not* some similarities, especially given the lack of objective criteria and specific guidelines for determining what constitutes a match between a modern birthmark and a historical case.

These are just a few logical and empirical problems with reincarnation, which is more deeply and thoroughly considered by the philosopher Paul Edwards in a book-length analysis, *Reincarnation: A Critical Examination*—still the best work on the topic. Edwards asks, rhetorically, "whether the belief in reincarnation should be regarded as false or as conceptually incoherent" and answers that he's inclined toward the latter view for reasons that become apparent the moment one seriously considers what's being proposed in reincarnation. When someone dies, his disembodied mind and/or nonphysical body (whatever that means) still somehow "retains memories of life on earth as well as some of his characteristic skills and traits; after a period varying from a few months to hundreds of years, this pure mind or nonphysical body, which lacks not only a brain but also any physical sense-organs, picks out a suitable woman on earth as its mother in the next incarnation, invades this woman's womb at the moment of conception of a new embryo, and unites with it to form a full-fledged human being." More problematic still, most of these reincarnated souls seem to prefer "to enter the wombs of mothers in poor and over-populated countries where their lives are likely to be wretched."[37] When you put it that way, the very idea of reincarnation is barking mad.

Nevertheless, as scientific skeptics we cannot simply ignore or dismiss the claims of supposed evidence in favor of reincarnation, however incoherent the idea may seem, so I shall consider reincarnation seriously now through one of the most famous examples I have encountered (I met the family on a television talk show). The case is that of a young boy named James Leininger, whose parents believe he is the reincarnation of a

World War II fighter pilot, the story of which is presented in their book titled *Soul Survivor*.[38] From a very young age, around two we are told, little James loved to play with airplanes. This is not so unusual for young boys, of course, but in time he began to have nightmares, and as his mother recalls, she'd wake him up and he would say things like, "Airplane crash on fire, little man can't get out." He made drawings of a burning, crashing plane and gave enough specifics about it that enabled his parents, through Google and archival searches, to link it to a twenty-one-year-old U.S. Navy fighter named James M. Huston Jr., who was shot down and killed by Japanese artillery in World War II during the battle for Iwo Jima. They determined that their boy was the reincarnation of this dead pilot.

How did the Leiningers derive this conclusion? By the time I met him, James was eleven and said he had only the vaguest memory of any of this. Tellingly, when pushed by the host, his mother, Andrea, confessed that "it wasn't like he had a cognitive memory," and that "these memories weren't active in his mind. It was just—it was usually a trigger or something that would happen or he would see or smell or hear something. And then he would just come out with this little piece of information and that was it." In fact, Andrea went on to admit that after this brief period in her son's life the memories were "pretty much gone forever." As she told us, "There was probably only three or five instances where we were able to sit down and question him and ask him questions. The rest of the time when we tried to do that, if he didn't initiate that conversation, he didn't seem to know what we were talking about. It was a very interesting phenomenon."[39]

Interesting indeed, given the aforementioned problem of false memories planted in people's minds, because if it wasn't little James who was driving this belief in the reincarnation of a World War II pilot, who was it? It began with Andrea's mother (James's grandmother), Barbara Scoggin, who suggested that James's nightmares might be the result of a negative past life memory. The seed planted, Barbara then suggested that Andrea consult one Carol Bowman, a reincarnation counselor and past lives regression therapist who guided the boy to "recover" more details about the plane and the deadly incident (she also wrote the foreword to the parents' book). "When we are dreaming, our conscious minds are not filtering material as when we are in a waking state, so unconscious material, including past life memories, emerge," Bowman told Andrea.

"It is not uncommon for young children to dream of their previous lives."[40] After rich details about the plane and fatal incident were reconstructed primarily by Bowman (with the assistance of Bruce and Andrea), the Leiningers wrote to James Huston's surviving sister, Anne Barron, who responded, "The child was so convincing in coming up with all the things that there is no way in the world he could know" (without little James being the reincarnation of her brother).[41]

We will examine those details below, but first, with regard to the population problem and personal identity, we might inquire where James Huston's soul was before little James was born. Was it occupying someone else's body, or floating around in the equivalent of a soul purgatory awaiting reimplantation? And what happened to James Leininger's soul while his body was occupied by James Huston's soul? Was Leininger's original soul floating around in some soul limbo awaiting a new incarnation or return to his original body after Huston's soul vacated it? Or does Leininger now have two souls, or even more from the past? According to Bowman, "As I see it, a part of James Huston's consciousness survived death and is a part of James Leininger's soul consciousness. The present incarnation is not a carbon copy of the last, but contains aspects of James Huston's personality and experience."[42] Which aspects, and why those? How does this work, exactly? We know how and where memories are stored in the brain via synaptic connections. How does the soul of someone else enter the brain and rewire the memory synapses? Does it do it by turning genes on and off, altering protein chain sequences, and repositioning synapses along the dendritic spines of neurons and their connections to other neurons, as happens when new memories are formed?

Second, why would James Huston return to the present day inside the body of James Leininger? "He came back because he wasn't finished with something," Bruce Leininger speculated.[43] More generally, says Bowman, "If a soul reincarnates with 'unfinished business,' or dies a traumatic death, these memories are more likely to carry over into another life. In James's case, he died a traumatic death as a young man. There was still much emotion and energy that may have propelled these memories forward."[44] If that were so, would this motive not apply to the millions of other young men whose lives were cut short in the deadliest war in human history? In the battle for Iwo Jima alone, 6,821 Americans were killed, every one of them with a long, full life ahead of him cut short and with, presumably, unfinished business. Did they all also inhabit the bodies of future children?

Third, how unusual is it for a young boy to dream about being a pilot? When I was a boy I fantasized about many such roles as I built model planes and ships and played war board games. I relived the World War I naval battle of Jutland and reenacted the World War II battle of Midway, with tiny ships and planes in my bedroom that spilled out into the hallway. Such is the rich fantasy life of youngsters. James's parents argue that their son exhibited characteristics that go far beyond those of a normal boy's fantasy, but who is to say what "normal" is in such cases? And by their own account they fueled James's imagination when they took him to the Cavanaugh Flight Museum in Addison, Texas, when he was only twenty months old. The boy, they said, was mesmerized by the planes, particularly those in the Second World War section of the museum where they wandered about for almost three hours. In that museum, on prominent display, is a Corsair, the plane James later mentioned from his dreams as the one Huston flew. On the way out of the museum Bruce purchased a videotape of the Navy Blue Angels planes, which he said little James practically wore out in repetitive viewing. By the parents' own account it was only *after* this trip that the nightmares about and waking comments on the World War II incident began. That timeline strongly suggests a causal vector of childhood fantasy, not reincarnation.

Admittedly, the claimed details about this case, such as a World War II pilot named James, the type of plane he flew (a Corsair), the ship from which it was launched (the aircraft carrier *Natoma*), the name of a fellow pilot (Jack Larson), drawings of a plane crashing, signing the sketches "James 3" (James Huston Jr. would have been James 2), and the chilling nightmare proclamation "Airplane crash on fire, little man can't get out," sound compelling upon first hearing. But when we consider that the boy's experiences, nightmares, and fantasies that resulted in this apparently coherent narrative were constructed only *after* the trip to the World War II museum that featured a Corsair plane, *after* the grandmother suggested past lives as an explanation, *after* the reincarnation therapist was consulted and engaged the boy in guided fantasy, *after* the father read to the boy books about World War II fighter planes, *after* the parents bought him toy planes, and *after* the parents became less skeptical and began to look for evidence to fit the reincarnation scenario, a more likely explanation emerges. And remember that James Leininger was three years old when he signed his name "James 3," so a signatory of age makes more sense than that a three-year-old somehow deduced that James Huston Jr. would have been "James 2." And the boy never mentioned the last name of the pilot he is allegedly reincarnating—"Huston"—only the first name James, which was his own and quite a common name at that.

Furthermore, it turns out that the plane James Huston was flying when he was shot down was a "Wildcat," not a "Corsair,"[45] and the ship he said Huston flew from, the *Natoma*, was actually named the *Natoma Bay*. That sounds as if I'm picking nits, but perhaps the *Natoma Bay* was pictured in one of those World War II books James's father read to him, and the unusual part of the name stuck. Finally, it is important to keep in mind that we have only the word of the parents for any of these claims, given that James Leininger himself says he no longer has reliable memories of any of this.[46]

———

A SCIENTIFIC UNDERSTANDING of NDEs and reincarnation such as I have offered here—situating them in natural explanations sufficient enough to forestall any need for invoking the supernatural—is not meant to take away from the power of the experience as seemingly real, as emotionally salient, or as transforming and life-changing. Perhaps belief in them is a way of dealing with the difficulties of life . . . and death. It is a self-transformation myth writ large. Resurrected. Born again. *Heavens on earth*.

CHAPTER 6

EVIDENCE FOR THE AFTERLIFE

Anomalous Psychological Experiences
and Talking to the Dead

> My parents died years ago. I was very close to them . . . I long to believe that their essence, their personalities, what I loved so much about them, are—really and truly—still in existence somewhere. Plainly, there's something within me that's ready to believe in life after death. And it's not the least bit interested in whether there's any sober evidence for it.
>
> —Carl Sagan, *The Demon-Haunted World*, 1996[1]

For a quarter of a century I have investigated and attempted to explain anomalous psychological experiences, and I have written about a few of my own, such as the sensation of having been abducted by aliens (caused by extreme fatigue and sleep deprivation during the three-thousand-mile nonstop transcontinental bicycle Race Across America), hallucinating inside a sensory deprivation tank (no light, no sound, floating in body-temperature warm water, one's mind begins to wander), and having an out-of-body experience while my temporal lobes were stimulated with electromagnetic fields (inside the lab of Michael Persinger while wearing his "God helmet").[2] Most people interpret such occurrences as evidence of the paranormal, the supernatural, the afterlife, or even God. For example, consider this remarkable anecdote recounted by the writer Mark Twain after a dream he had about his brother Henry, with whom he was working on a riverboat docked in St. Louis:

In the morning, when I awoke I had been dreaming, and the dream was so vivid, so like reality, that it deceived me, and I thought it was real. In the dream I had seen Henry a corpse. He lay in a metallic burial-case. He was dressed in a suit of my clothing, and on his breast lay a great bouquet of flowers, mainly white roses, with a red rose in the center.

In point of fact, several weeks later Twain's brother Henry died from an overdose of opium administered to him after he was burned in a boiler explosion aboard the ship. Twain recalled:

When I came back and entered the dead-room Henry lay in that open case, and he was dressed in a suit of my clothing. He had borrowed it without my knowledge during our last sojourn in St. Louis; and I recognized instantly that my dream of several weeks before was here exactly reproduced, so far as these details went—and I think I missed one detail; but that one was immediately supplied, for just then an elderly lady entered the place with a large bouquet consisting mainly of white roses, and in the center of it was a red rose, and she laid it on his breast.[3]

Twain himself was ambivalent as to how he should interpret this anomalous experience, admitting that he had told and retold the story so many times before committing it to paper that it was entirely possible he had festooned it with details to make a prosaic explanation seem more unlikely. Plus, each of us has about 5 dreams per night, or 1,825 dreams per year. If we remember only a tenth of our dreams, then we recall 182.5 dreams per year. Using a rounded-off figure of 300 million Americans able to remember their dreams, this generates a total of 54.7 billion remembered dreams per year. Sociologists who study social networking estimate that each of us knows about 150 people fairly well, for a total network social grid of 45 billion personal relationship connections. With an annual death rate of 2.4 million Americans per year at all ages and from all causes, it is inevitable that some of those 54.7 billion remembered dreams will be about some of these 2.4 million deaths among the 300 million Americans and their 45 billion relationship connections. In fact, it would be a *miracle* if some death premonition dreams did not come true.

Nevertheless, there are those who interpret such anomalous experiences as veridical of the afterlife scenario, one of whom is the Rice

University religious studies professor Jeffrey Kripal, who uses the Twain death dream as an example of what he calls "traumatic transcendence," or "a visionary warping of space and time effected by the gravity of intense human suffering." Kripal embraces the early Victorian characterization of such dreams as "veridical hallucinations," or "hallucinations corresponding to real events," and notes that "individuals have been seeing dead loved ones (or loved ones about to die at a distance) for millennia, which suggests strongly that experiences like those of Twain . . . are very much a part of our world and not simply constructed by culture." Kripal concludes that "such comparisons are deeply suspect these days, mostly because they end up suggesting something at work in history that is not strictly materialist—like a mind that knows what is going to happen before it happens, or a departed soul that appears to his sleeping wife." He claims that in fact "we are swimming in a sea of such stories, if only we could recognize our situation. We do not know how many such stories there might be, much less what they might mean. We do not know because we have never really tried to find out."[4]

AFTERLIFE EXPERIMENTS

There is in fact a rich literature of anomalous psychological experiences that I and many others in the scientific community have amassed in books, magazine articles, documentary films, blogs, podcasts, and other media, such as those compiled by Graham Reed in *The Psychology of Anomalous Experience*,[5] by Leonard Zusne and Warren Jones in *Anomalistic Psychology*,[6] and especially by the psychologists Etzel Cardena, Steven Lynn, and Stanley Krippner in their comprehensive *Varieties of Anomalous Experience*. From a scientist's perspective, the latter authors contend that an anomalous experience is "an uncommon experience (e.g., synesthesia) or one that, although it may be experienced by a substantial amount of the population (e.g., experiences interpreted as telepathic), is believed to deviate from ordinary experience or from the usually accepted explanations of reality."[7]

In other words, weird things happen and curious minds want to know why. In addition to interpreting such anomalous experiences as evidence of the numinous, some go so far as to take them as proof of life after death, most notably those who call themselves psychic mediums, such as James Van Praagh, John Edward, and Rosemary Altea, who claim that they can talk to the dead—in television studios and hotel conference rooms, no less, on cue at the appointed time for studio audiences and paying customers seeking to make contact with their lost loved ones. They even claim to know what heaven is like. In his book *Growing Up in Heaven*, for example, Van Praagh offers this description of what deceased children have to look forward to:

> Besides attending school, children entertain themselves with a wide range of activities like sports, games, and handicrafts. They enjoy swimming, cycling, gardening, baseball, woodcarving, sailing, and anything else they desire . . . Some become members of organizations similar to the Boy Scouts and Girl Scouts that teach leadership skills, teamwork, arts and crafts, conservation, and camping.[8]

Cycling in heaven? I could go for that, but does anyone take this seriously? They do, especially bereaved parents, who are eager to hear such messages as these that Van Praagh channeled: "You were a great dad—the best. He says he loves you both very much. He wants you to know that. He says he picked you especially as his parents and wants you to know that you will always be together." Sensing guilt from another parent, Van Praagh reassured, "He wants you to forgive yourself. You did nothing wrong. Do you understand? *You're not to blame*, he is saying."[9]

Such vapid pronouncements need no extensive rebuttal here, as in several of my previous books and in numerous articles I have exposed these alleged psychics who claim to communicate with the deceased as nothing more than charlatans employing cold reading techniques—methods by which you read someone "cold" in the sense that you've never met them. And, as I like to say, anyone can talk to the dead; it's getting the dead to talk back that is the hard part. Making it *appear* that the dead are communicating with the living, however, is remarkably easy, as I discovered when I spent a day doing readings of random people I'd never met for Bill Nye's television series *The Eyes of Nye*. Despite knowing nothing about the dozen or so people I did readings on, most confirmed that I was eerily accurate.[10] Such is the power of belief, coupled with some surefire techniques from Ian Rowland's *The Full Facts Book of Cold Reading*.[11]

In brief, says Rowland, when doing a psychic reading, adopt a soft voice, a calm demeanor, and sympathetic and nonconfrontational body language: a pleasant smile, constant eye contact, head tilted to one side while listening, and facing the subject with legs together (not crossed) and arms unfolded. Ask a lot of questions, such as "Who is the person who has passed over that you want to try to contact today?" And: "Is this making sense to you?" "This is significant to you, isn't it?" "You can connect with this, can't you?" "So who might this refer to, please?" And so forth. You can claim connection to the dead through details only they would seem to know but are in fact quite common. For example, says Rowland, make references to: a scar on a knee; the number 2 in a home address; a childhood accident involving water; clothing never worn; photos of loved ones in a purse or wallet; wore hair long as a child; one earring with a missing match. And so on. If done with confidence and a little bit of practice, it is quite easy to convince almost anyone that you are able to talk to the dead.

Unfortunately, not everyone is so skeptical of this form of evidence for an afterlife, chief among them being Gary Schwartz of the University of Arizona who, in his assuredly titled book *The Afterlife Experiments: Breakthrough Scientific Evidence of Life After Death*, claims that by studying psychic mediums during their supposed communications with the deceased under controlled conditions, he has proved that the afterlife is real.[12] In fact, as Marc Berard notes in his careful analysis of this research, the setup in Schwartz's lab for these readers was less than ideal.[13] In some conditions the psychics were given feedback by the sitters (the people getting the reading), while in others the sitters nodded affirmatively or negatively to a comment by the psychic while sitting behind a blind and the experimenter then answered for them, which is still feedback. When no feedback was provided the mediums' readings were less accurate. Schwartz was also willing to count as a hit a statement that was a miss for one sitter but a hit for another. That's not allowed in controlled experiments. In one bizarre twist, one of Schwartz's mediums inquired if the sitter's husband had died. Although it was her son who had passed away, it was mistakenly relayed to the psychic that it was the husband who died, and he proceeded to channel from the other side the still living spouse. When the woman's husband was later killed in an accident, Schwartz proclaimed the psychic reading to be *prophetic*. This is a postdiction by Schwartz, not a prediction by the psychic.

Further, Schwartz does not seem to understand the nature of cold reading, believing it to be largely the process of guessing, reading the body language of sitters, or garnering information on sitters ahead of time. This is not the basis of cold reading, which depends largely on making comments that are true for most people in most circumstances most of the time, and emphasizing the hits and downplaying the misses. For example, Schwartz makes much about the fact that several of the mediums he studied mentioned a dog in relation to the sitters. If offered in a general enough statement—they have a dog, they *had* a dog at some point, or they knew someone who has or had a dog—the hit rate is bound to be high. In the opinion of the estimable Ray Hyman, a psychologist who has performed and studied cold readings for half a century:

> Again and again, Schwartz argues that the readings given by his star mediums differ greatly from cold readings. He provides samples of readings throughout the book. Although these samples were obviously selected because, in his opinion, they represent mediumship at its best, every one of them strikes me as no different in kind from those of any run-of-the-mill psychic reader and as completely consistent with cold readings.

After studying Schwartz's experimental setup carefully, Hyman, who is also an expert in experimental design in psychological research, identified ten flaws in the afterlife experiments: (1) inappropriate control comparisons; (2) inadequate precautions against fraud and sensory leakage; (3) reliance on nonstandardized, untested dependent variables; (4) failure to use double-blind procedures; (5) inadequate "blinding" even in what he calls "single blind" experiments; (6) failure to independently check on facts the sitters endorsed as true; (7) use of plausibility arguments to substitute for actual controls; (8) the confusion of exploratory with confirmatory findings; (9) the calculation of conditional probabilities that are inappropriate and grossly misleading; (10) creating nonfalsifiable outcomes by reinterpreting failures as successes.[14] Far from proof of the afterlife, psychic mediums demonstrate the power of cognitive dissonance to maintain belief in the teeth of counterevidence.[15]

FIND THE STRANGEST THING AND THEN EXPLORE IT

More skeptical investigators explore the possibilities of finding evidence for or making contact with an afterlife through such technologies as EEG and fMRI brain scans of meditators, mediums, and spiritualists of various stripes, most notably by the neuroscientist Andrew Newberg, whose book titles convey his optimism that such a numinous world exists: *Why God Won't Go Away*, *Born to Believe*, and *How God Changes Your Brain*.[16] In one such study, for example, Newberg and his colleagues scanned the brains of ten psychographers, or mediums who allegedly channel messages from dead spirits through a technique called automatic writing, in which their hands are supposedly directed by spirits or forces from beyond to produce messages. They found that experienced psychographers wrote more complex messages but showed lower levels of activity in the parts of the brain associated with writing and related cognitive processes, suggesting that it was the spirit doing the writing and not the medium (through faking or role playing).[17] Although Newberg and his colleagues are cautious in the wording of this spiritual suggestion, the implication is there nonetheless.[18] But I think a more cautious interpretation of the data suggests that experienced mediums and psychographers have been practicing their craft for so long that it has become automated—it's called automatic writing, after all—and thereby imposes a lighter cognitive load than that of less experienced subjects, much like the muscle memory of experienced athletes, musicians, and artists, who report being in a state of flow when they need not even think about what they're doing.

Although this line of research is admittedly more sophisticated than that involving the automatic "slate writing" of the nineteenth century, exposed as a magic trick by the magician Harry Houdini (and others),[19] the implications are nevertheless spiritual in nature: that the dead can communicate with us from the other side. Can they?

I am skeptical, but open-minded—not only because that is the approach any scientist should take when confronting an unexplained mystery, but also because of my own experiences with the anomalous. Most dramatically, in early 2014 I had a jarring experience for which I have no explanation. Later that year I wrote about it in my *Scientific American* column. Entitled "Infrequencies," the article has generated more mail than any of my other columns since I began writing it in 2001.[20] In brief, my fiancée at the time, Jennifer Graf (now my wife), moved to Southern California from Köln, Germany, bringing with her a 1978 Phillips 070 transistor radio that had belonged to her late grandfather Walter, a surrogate father figure, as she was raised by a single mom. She had fond memories of listening to music with him through that radio, so I did my best to resurrect it, without success. After putting in new batteries and leaving the power switch in the "on" position, we gave up and tossed it in a desk drawer in our bedroom, where it lay dormant for months. During a quiet moment after our vows at a small wedding ceremony at our home, Jennifer was feeling sad being so far from her family and friends and wishing she had some connection to loved ones—most notably her mother and her grandfather—with whom to share this special occasion. We left my family to find a quiet moment alone elsewhere in the house when we heard music emanating from the bedroom, which turned out to be a love song playing on that radio in the desk drawer. It was a spine-tingling experience. It could have been tuned to any station—or more likely between stations producing nothing but static—but it was perfectly tuned to a station playing music suited to the occasion. And it could have come on at any point in the months before or after the wedding ceremony, but it happened at the very moment Jennifer most needed that connection. The radio played similar music for the rest of the evening but went quiescent the next day. It's been silent ever since, despite my repeated attempts to revive it.

Ever since that column appeared, I've been deluged with letters. A few grumpy skeptics chided me for lowering my skeptical shields, most notably for my closing line: "And if we are to take seriously the scientific credo to keep an open mind and remain agnostic when the evidence is indecisive or the riddle unsolved, we should not shut the doors of perception when they may be opened to us to marvel in the mysterious." It was, admittedly, a little poetically playful in referencing the title of Aldous Huxley's book *The Doors of Perception*, which I qualified by noting: "The emotional interpretations of such anomalous events grant them significance regardless of their causal account."

A number of believers sent encouraging notes, not all of which I understood, such as this sentiment from a psychologist: "The central importance of latent, neglected shared spiritual capabilities was indeed a wedding blessing, eloquently and vividly enacted, resulting in very valuable sharing for a world culture remarkably crippled in appreciation of actual multidimensional reality." Does three-dimensionality count? A neurophysiologist imagined what the implications would be if no natural explanation were forthcoming for my anomalous event. "Should consciousness survive the death of the brain, there are exciting implications for the role of consciousness in the living brain." Indeed there is, but a lack of causal explanation for my story does not imply this.

A geologist wrote to suggest that "There are many explanations that can be posited; I would favor solar flares or the geoparticles of Holub and Smrz [authors of a paper that some claim proves that nanoparticles between neurons may allow for quantum fields to influence other brains], but rather than seek one, this coincidental occurrence should be enjoyed in the supernatural or paranormal vein as it was meant to be . . . simply a blessing for a long and happy union." I agree, but without the supernatural or paranormal vein in the rock.

Most of the correspondence I received, however, was from people recounting their own anomalous experiences that had deep personal meaning for them, some pages long in rich detail. One woman told me the story of her rare blue opal pendant that she wore 24/7 for fifteen years, until her husband swiped it out of spite during their divorce. She felt so bad that while on vacation in Bali she had a jeweler create a simulacrum of it, which led her to establish and subsequently to enjoy a successful jewelry business for many years. One day, a woman named Lucy came into her store and they got to talking about the lost opal pendant, which Lucy suddenly realized that she now owned. "In 1990 her best friend was dating a guy who was going through a divorce and he had given it to her. Her friend never felt comfortable wearing it so she offered it to Lucy. Lucy accepted, and wore it the following weekend on her wedding day. Soon after, she discovered her new husband had a girlfriend, and she never wore the opal again, thinking it might be bad luck. It remained in her drawer for fifteen years. When I asked why she hadn't sold it (it was now extremely valuable), she said 'I tried to—every time I went to get it out of the drawer to have it appraised, something happened to distract me. Phone calls, dogs fighting, package deliveries—I tried many times, but never succeeded. Now I know why—it wanted to come back to you!'" This woman's sister, whom she characterized as a "medical intuitive and remote healer," called this story "epic synchronicity." She described it as "fantastic and statistically improbable, but it *is* explainable."

I agree, but what is the explanation for this, or for any of such highly improbable events? And what do they mean? For Jennifer and me, our anomalous experience was the propitious *timing* of the radio's revival—at the moment she was thinking about family—that made it such an emotionally salient event, enabling her to feel as if her beloved grandfather was there with us, sharing in our commitment. Does this constitute proof of life after death? No. As I wrote, "such anecdotes do not constitute scientific evidence that the dead survive or that they can communicate with us via electronic equipment."

The reason is that in science it isn't enough to just compile anecdotes in support of a preferred belief. After all, who wouldn't want to know that we and our loved ones survive bodily death and live for eternity elsewhere? As Carl Sagan suggested in the epigraph for this chapter, it's a powerful desire we all have to be connected to our lost loved ones. "Sometimes I dream that I'm talking to my parents, and suddenly—still immersed in the dreamwork—I'm seized by the over-powering realization that they didn't really die, that it's all been some kind of horrible mistake," Sagan elaborated, noting that this also explains the hallucinations he has experienced:

> Probably a dozen times since their deaths I've heard my mother or father, in a conversational tone of voice, call my name. Of course they called to me often during my life with them—to do a chore, to remind me of a responsibility, to come to dinner, to engage in conversation, to hear about an event of the day. I still miss them so much that it doesn't seem at all strange that my brain will occasionally retrieve a lucid recollection of their voices.[21]

Such longings make us all subject to a number of cognitive biases, most notably the *confirmation bias* in which we look for and find confirming evidence and ignore disconfirming evidence. We remember highly unusual coincidences that have deep meaning for us and forget all the countless meaningless coincidences that flow past our senses every day. Then there is the *Law of Large Numbers*: with seven billion people having, say, ten experiences a day of any kind, even million-to-one odds will result in seventy thousand coincidences a day. It would be a wonder if at least a few of those events did not get remembered, recounted, reported, and recorded somewhere, leaving us with a legacy of frequently infrequent anomalies. Add to this the *hindsight bias*, in which we are impressed by the improbability of an event after the fact; in science, however, we should be impressed only by events whose occurrences were predicted in advance. And don't forget the *recall bias*, in which we remember things that happened differently depending on what we currently believe, retrieving from memory circumstances that favor the now preferred interpretation of the event in question. Then there is the matter of what *didn't happen* that would have been equally spine-tingling in emotional impact on that day, or some other important day, and in my case I can't think of any because they didn't happen. Finally, just because I can't explain something doesn't mean it is inexplicable by science. The *argument from ignorance* (it must be true because it has not been proven false) or what Richard Dawkins calls the *argument from personal incredulity* (because I cannot imagine a natural explanation then there cannot be one) doesn't hold water on the scientific seas.

Such problems were highlighted by the evolutionary biologist Jerry Coyne in his critique of Jeffrey Kripal's article about anomalous experiences with which we began this chapter. Kripal, Coyne begins, "doesn't note the far more frequent instances of 'precognition' that *don't* come true and aren't reported, nor the notorious tricks that human memory can play." He calls attention to the fact that "there are plenty of cases, for example, of humans selectively editing their memories in retrospect to conform to what they—or others—want to believe. This alone makes peculiar anecdotes like Twain's deeply suspect, or at least in need of scientific confirmation." Coyne also adroitly notes that "Kripal doesn't tell us why the great majority of people who die or suffer in the absence of their loved ones don't send telepathic messages conveying their distress. Do they lack the right kind of transmitter? And why aren't we all receiving numerous psychic signals which must, after all, be crisscrossing the atmosphere like radio waves?"[22] In other words, Kripal has fallen for a host of cognitive biases in support of his thesis that there exists a spiritual world outside the material one, but he has not told us what it would take to falsify his hypothesis, and he has certainly not tried to do so. The falsification process is at the heart of science for these many reasons.

As for plausible explanations for the broken radio experience of my wife and me, Deepak Chopra suggested that "the radio coming on and off almost certainly has a mechanical explanation (a change in humidity, a speck of dust falling off a rusty wire, etc.). What is uncanny is the

timing and emotional significance to those participating in the experience. The two of you falling in love is part of the synchronicity!" Agreed,
and I can well imagine some electrical glitch, a particle of dust, an electromagnetic fluctuation from the batteries—something in the natural
world—caused the radio to come to life. Why it would happen at that
particular moment, and be perfectly tuned to a station playing love
songs, and be loud enough to hear out of the desk drawer, is what made
the event stand out for us.

A psychologist named Michael Jawer wrote to offer his explanation
"that strong and underlying feelings are central to anomalous happen-
ings." His approach "doesn't rely on barely-understood quantum woo,"
he cautioned, "but assesses the way feelings work within our biology
and physiology and the way emotions knit human beings together."
That certainly sounds reasonable, although how emotional energy could
be transmitted from inside a body (or from the other side) into, say, a
radio is not clear. But I appreciated the close of his letter in which he
quoted the late physicist John Wheeler: "In any field, find the strangest
thing and then explore it."[23]

REVELING IN THE MYSTERY

Exploring the strangest thing is precisely what the eminent Caltech
physicist Kip Thorne did in the blockbuster film *Interstellar*, for which
he was the scientific consultant. In order to save humanity from imminent extinction on Earth as a result of catastrophic climatic changes,
Matthew McConaughey's character, Cooper, has to find a suitable
planet on which to live by passing through a wormhole to another galaxy.
In order to return, however, he must slingshot around a black hole,
thereby causing a massive time dilation relative to his young daughter,
Murph, back home on Earth (one hour near the black hole equals seven
years on Earth), such that by the time he returns, Murph is much older
than he. In the interim, in order to get the humans off Earth, Cooper
needs to transmit information to his now adult scientist daughter about
quantum fluctuations from the singularity inside the black hole. To do
so, Cooper uses a *tesseract*—a multidimensional analog of a cube in
which time appears as an extra spatial dimension that includes portals
into the daughter's childhood bedroom at a moment when (earlier in

the film) Murph thought she experienced ghosts and poltergeists myste-riously pushing books off her bookshelf.

These experiences turned out to be her father from the future reach-ing back in time through extra dimensions via gravitational waves to get her attention, after which he transmits critical data via Morse code dots and dashes on the second hand of the watch he left her in order to facilitate actions needed to save humanity. It's a farfetched plot, but according to Thorne in his companion book to the film, *The Science of Interstellar*, it's all grounded in natural law and forces. "When falling into and through the tesseract, Cooper truly does travel backward rela-tive to our brane's [universe's] time, from the era when Murph is an old woman to the era when she is ten years old," Thorne explains. "He does this in the sense that, looking at Murph in the tesseract bedrooms, he sees her ten years old. And he can move forward and backward relative to our brane's time (the bedroom's time) in the sense that he can look at Murph at various bedroom times by choosing which bed-room to look into."[24]

This is another way of saying that there is no such thing as the supernatural or the paranormal. There is just the natural and the nor-mal and mysteries we have yet to solve with natural and normal expla-nations. If it turns out, say, that Jennifer's grandfather Walter exists in a multidimensional tesseract in which he can see her at all times of her life simultaneously, and used gravitational waves near a black hole or a wormhole to turn on his old radio for his granddaughter at that partic-ular time when she most needed it, that would be fully explicable by physical laws and forces as we understand them. It would not be ESP or Psi or anything of the paranormal or supernatural sort. It would just be a deeper understanding of physics. If there is an afterlife, its existence would be explained by something along these lines, although here we don't even know what we don't know, so this is pure speculation.

Until such time when science can explain even the most spectacularly unlikely events, what should we do with such stories? Enjoy them. Appreciate their emotional significance. Embrace the mystery. What we do not need to do is fill in the explanatory gaps with gods or any such preternatural forces. We can't explain everything, and it's always okay to say "I don't know" and leave it at that until a natural explanation presents itself. Until then, revel in the mystery and drink in the unknown. It is where science and wonder meet.

CHAPTER 7

SOUL STUFF

Identity, Replication, and Resurrection

The real question of life after death isn't whether or not it
exists, but even if it does what problem this really solves.

—Ludwig Wittgenstein, *Tractatus Logico Philosophicus*, 1921

In the episode of *Star Trek: The Next Generation* titled "Second Chances,"
Commander William Riker of the Starship *Enterprise* beams down to
a planet to retrieve data from a research station he visited eight years
before when he was a lieutenant on board the Starship *Potemkin*. There
he discovers an exact duplicate of himself, the product of the *Potemkin*'s transporter beam accidentally being split into two and material-
izing a second Riker after the original beamed back to the ship. Lieutenant
Riker remained stranded on the planet while his other self continued his
life trajectory in Starfleet, where he moved up the ranks to become
Commander Riker of the *Enterprise*. DNA and brain scans of the two
Rikers reveal that they are genetically identical and neurologically indistinguishable. They are true duplicates. Lieutenant Riker's lover before the
transporter mishap, Counselor Deanna Troi, is no longer romantically
involved with Commander Riker on the *Enterprise*, and much of the
episode plays out the awkwardness of experiencing and reexperiencing
the breakup for the two Rikers and Troi. In the end Lieutenant Riker is
assigned a post on a different ship and adopts his middle name as his first
in order to distinguish himself from his newfound twin.[1]

Were the two Rikers two different people or duplicates of one person? If they were true duplicates, did they subsequently become two different persons the moment they started leading separate lives and forming new memories and identities? This is the essence of the *identity problem* and it is vital to solve for all resurrection scenarios, both religious and scientific.

THE IDENTITY PROBLEM

The identity problem was first articulated by the ancient Greek scholar Plutarch in his thought experiment known as the "ship of Theseus." According to the myth, Poseidon's son Theseus sailed to Crete where he slew the half-man/half-bull Minotaur monster. After his triumphant return to Athens, Theseus' ship was preserved in memoriam. As the vessel aged, however, the decaying wood was gradually replaced with new timber until eventually the entire ship was made of different material. Was it still Theseus' ship?

The answer depends on how you define the true identity of a thing—as the pattern or the material.[2] If Theseus' ship is represented by the pattern, then replacing all its lumber does not alter its identity. If the ship's distinctiveness is held in the material of which it is made, however, or in some combination of pattern and material, then altering the physical structure changes the identity in some manner. But how much would need to be exchanged before it was no longer the same "thing," no longer Theseus' ship?

Take our bodies. In addition to the replacement of atoms, molecules, cells, tissues, and organs every few years, there is a huge number of "foreign" cells inside us that contain no human DNA or RNA—bacteria that produce chemicals that enable our bodies to process the energy and nutrients in the food we eat, others that boost immunity, and still others whose function remains mysterious.[3] More identity-shattering still, it appears that the complex eukaryotic cells of which we are made evolved billions of years ago from much simpler prokaryotic cells in a process the evolutionary biologist Lynn Margulis calls *symbiogenesis*— the cooperative union of primitive simple prokaryotic cells into modern complex eukaryotic cells.[4] The membrane-bound mitochondria organelles inside our cells that are so vital to the processing of energy, for example, have their own DNA different from that found in the nucleus of the cell (the famous mitochondrial DNA from which our genetic heritage can be traced over millions of years). It is now commonly believed that around 1.5 billion years ago some of these free-living bacteria (prokaryotes) symbiotically cooperated to form the more complex eukaryotic cells that make up modern organisms like us. So if you go back far enough in evolutionary time, even the contents of our cells are foreign.

And yet we don't feel like a collection of other organisms. We feel like a whole self. The pattern of biological information coded in our genome, and the neural synaptic arrays recorded in our brain's connectome, assure this continuity of essence. You are still you across space and time, even though the material making you up changes. Our sense of identity remains intact despite the exchange of body stuff, so our uniqueness appears to be ingrained in the pattern more than the material.

By this analysis, would a duplicate of you also be you, even if it meant that there is more than one of you? In principle, yes, as long as each of the duplicates feels like an autonomous person. This is why, in addition to the *pattern* and the *material* of identity, there is an additional component: *personal perspective,* or Point of View (POV). The neuroscientist Kenneth Hayworth, whom we will meet in the next chapter, calls this the POVself, which he contrasts with the MEMself, or the complete set of your memories. Every self-contained sentient being—by which I mean the capacity to be *emotive, perceptive, sensitive, responsive,* and *conscious*—is a POVself because it has a personal perspective, and that is what makes each person an autonomous identity. By this definition, in the *Star Trek* scenario each Riker is his own POVself, even though at the moment of replication each had an identical MEMself, in the same way that identical twins are two persons, both psychologically and legally.

However, there is an important proviso to the duplication scenario, and that is that the moment you and your duplicate MEMself (or identical twin) begin to lead separate lives, you are not only separate POVselves; you are also separate MEMselves—not just because of the different perspectives but also because of the different experiences you have, forming distinct memories, personalities, and all the rest that goes into the makeup of your pattern of information. Research shows, for example, that in the case of twins, this uniqueness of experience begins in the womb, given the different positions, sounds, nutrients, and all the rest to which each fetus is uniquely exposed. And even identical twins are not perfectly genetically identical.[5] A study by the geneticist Michael Lodato and his colleagues discovered that each of the billions of neurons in a brain may contain up to fifteen hundred unique mutations that arise during cell division and duplication, from environmental factors (radiation, chemicals), and from switching on and off, depending on the environment and circumstances, when they're needed and when they're not, which in turn causes neurons to grow new synaptic connections in response to any number of conditions, such as learning something new, memorizing information, solving problems, or encountering new people or social arrangements.[6] When those genes unfold to do their thing, they may undergo mutations, making them unique from other such neurons.

The dogma that every cell in your body contains identical DNA—your DNA—is also being challenged. A study by the geneticist Michael McConnell and his colleagues, for example, indicates that up to 40 percent of the neurons in your brain contain large chunks of DNA that have either been duplicated or deleted, and others that have "jumped" to other host neurons, and that "a subset of neurons have highly aberrant genomes marked by multiple alterations."[7] "The assumption has always been that the genomes within every individual are identical," McConnell explained in discussing the implications of his research. "Now that we know that assumption is false, it's forced a rethink."[8] Schizophrenia, for example, runs in families and has thus been assumed to be heritable, and yet genes account for a small percentage of those who succumb to this mental disorder, which explains why most people in a family do not come down with the disorder even if one or two members do. The genetic sequences leading to this and other diseases may be unique to each patient's life history as it affects their genes, in addition to the genetic heritage with which they were born.

THE SOUL

The neurobiologist and philosopher Owen Flanagan summarizes the three primary characteristics of the soul:[9] the *unity of experience* (a sense of self or "I"), *personal identity* (the feeling of being the same person over the course of a lifetime), and *personal immortality* (the survival of

Figure 7-1. The Soul

William Blake's portrayal of the soul departing the body upon death captures
what most people believe to take place. An illustration from a series designed
by Blake for an edition of the poem "The Grave" by Robert Blair, engraved by
Louis Schiavonetti in 1813, titled *The Soul Hovering over the Body, Reluctantly
Parting with Life.* Courtesy of the Metropolitan Museum of Art.

death) (see William Blake's depiction of the soul in figure 7-1).[10] Polls
consistently show that between 70 and 96 percent of Americans believe
in a soul as so characterized.[11] The vast majority of people base such
belief on religious faith, but science tells us that all three of these characteristics are illusions.

Unity of Experience. There is no unified "self" that generates inter-
nally consistent and seamlessly coherent beliefs devoid of conflict. Instead,
we are a collection of distinct but interacting modules—or neural
networks—that are often at odds with one another. According to the
evolutionary psychologist Robert Kurzban, the brain evolved as a
modular, multitasking, problem-solving organ—a Swiss Army knife of
practical tools, in the old metaphor, or an app-loaded iPhone, in Kurzban's upgrade.[12] The module that leads us to crave sweet and fatty foods
in the short term, for example, is in conflict with the module that moni-

tors our body image and health in the long term. The module for cooperation is in conflict with the module for competition, as is the module
for truth telling with the module for lying. Of course, because the brain
does not sense itself operating, we are blissfully unaware of all these
networks running largely independently, so it feels as though there is a
unity of self.[13]

Personal Identity. Scientists estimate that in the course of your lifetime, most of the atoms in your body will be replaced by comparable
atoms—hydrogen atoms most rapidly (given that our bodies consist
of 72 percent water, which is two parts hydrogen and one part oxygen),
then heavier elements such as carbon, sodium, and potassium.[14] As
atoms are replaced, so, too, are molecules, cells, tissues, and organs, by
some estimates on average every seven to ten years. There is a wide
variation of the replacement process time, from a few days for the
epithelial cells that line the gut, to a few weeks for the epidermis skin
layer, to two months for red blood cells, to a year or two for liver cells,
to ten to fifteen years for bone and muscle.[15] So the belief that you are
the same material person you were years ago—or will be years from
now—is an illusion. At most, what stays the same is the pattern of
information, and even this changes over time.

Personal Immortality. We have already seen that there is no evidence
for an afterlife as proposed by religionists, but what about a scientific
immortality? Duplication is not an option for immortality unless there
is a continuity of self from one duplicate to the next. When you fall
asleep or go under general anesthesia for surgery, despite the disruption
in consciousness of several hours you still feel like yourself when you
wake up. How, exactly, would that happen if you were duplicated, replicated, resurrected, or uploaded? If a brain could be cryopreserved and
reawakened after, say, a thousand years, would it be the same as waking
up from a long sleep? Maybe. What about a brain whose connectome of
information is precisely recorded and uploaded into a computer? When
it is turned on, would the personal perspective of the person be in there?
Maybe not.

The identity problem confronts both religious and nonreligious
seekers of immortality. If you are religious and believe in the resurrection of the body or the soul in heaven, for example, how does God go
about the duplication or transformation process to ensure not just the
MEMself continues, but also the POVself? Is it your atoms and patterns

that are resurrected, or just the patterns? If both, and you are physically resurrected, does God reconstitute your body so it is no longer subject to disease and aging? If it is just the patterns of you that are resurrected, what is the platform that holds the information? Is there something in heaven that is the equivalent to a computer hard drive or the cloud? Is there some sort of heavenly quantum field that retains your thoughts and memories? When you die, will you "wake up" in heaven as if after a sleep? That is, if God is duplicating your MEMself, how does the POVself come with it? If you are not religious (or even if you are) and you hold out hope that one day scientists will be able to clone your body and duplicate your brain, or upload your mind into a computer, or create a virtual reality in which you come alive again, we face the same technological problems as God would in accomplishing the transference, particularly of your personal perspective, or POVself.

So our self is defined by our physical makeup, our patterns of information, our unique experiences, and our personal perspective. This makes us autonomous selves. This is the real you. This is your *soul*.

SOUL SEARCHING

Consider this thought experiment. I'm a serious cyclist who takes risks in the hazardous dense traffic of Southern California and on the treacherous descents of the local mountain roads. Because of this I have deposited a complete copy of my DNA with 23andMyClone, Inc., which has the technology to produce a perfect replica of me should I not survive a horrific crash. I have also contracted with MindCloud Computers to maintain a full backup of my mind in the cloud, and they remind me every evening to update the backup while I'm sleeping. It's like life insurance, only instead of giving my wife a large cash payoff upon my death, she gets a perfect copy of me with all my thoughts and memories. It is so perfect a replica of me that even my family and friends cannot tell the difference.

One day my wife gets a call from the California Highway Patrol that I've gone off the side of a cliff while descending the switchbacks from Mount Wilson and they found my mangled body at the bottom of a ravine. It appears I have died. She's upset (I hope) and calls 23andMyClone and orders a copy, which they had ready-made in case I needed any of his organs as I aged. Then she contacts MindCloud Computers and has them upload my mind into the clone. In a matter of days my wife has her husband back home and all is well. Except this: I was not actually dead. Thanks to my helmet I was only in a deep coma from the crash, and because it was snowing that day my body was cooled enough to preserve my tissues, including my brain, for several days. I wake up after a few days and decide to surprise my (presumably) grief-stricken wife at home. As I enter the house I hear voices coming from the bedroom. Puzzled, I quietly go upstairs into the master bedroom to find my wife in bed with . . . me!

Now, clearly, Michael II is not Michael I no matter how good the copy is, even if it were perfect, because *I'm* Michael Shermer, not that guy with my wife. That dude is just a copy and I'm not at all happy at what I see. Of course, Michael II would have the same feelings of love for my wife that I have, along with all of the rest of my thoughts and feelings and memories right up to the time of the last backup. After that, Michael II would immediately begin experiencing things and storing memories that I never had. Michael II is a copy of my brain and body, but without the continuity of the personal perspective that I experience when waking up from sleep or general anesthesia, it's not *me* continuing on with life. When 23andMyClone and MindCloud Computers constructs Michael II, my inner self does not wake up inside his head. There is no continuity between Michael I and Michael II. There is a discontinuous break, and so any theory of immortality, resurrection, or afterlife—religious or scientific—must close this gap.

A GOOGOL GOOGOLPLEX OF RESURRECTION COMPLEXITY

The sums involved in achieving immortality through the duplication or resurrection scenarios are not to be underestimated. There are around 85 billion neurons in a human brain, each with about a thousand synaptic links, for a total of 100 trillion connections to be accurately preserved and replicated. This is a staggering level of complexity, made all the more so by the additional glial cells in the brain, which provide support and insulation for neurons and can change the actions of firing neurons, so these cells had better be preserved as well in any duplication or resurrection scenario, just in case.[16] Estimates of the ratio of glial

cells to neurons in a brain vary from 1:1 to 10:1. If you're not a lightning calculator, that computes to a total brain cell count of somewhere
between 170 billion and 850 billion. Then factor in the hundreds or
thousands of synaptic connections between each of the 85 billion neurons, adding approximately 100 trillion synaptic connections for each
brain. That's not all. There are around 10 billion proteins per neuron,
which affect how memories are stored, plus the countless extracellular
molecules in between those tens of billions of brain cells.

These estimates are just for the brain and do not even include the
rest of the nervous system outside the skull—what neuroscientists call
the "embodied brain" or the "extended mind" and which many philosophers of mind believe is necessary for normal cognition. So you might
want to have this extended mind resurrected or uploaded along with
your mind. After all, you are not just your internal thoughts and emotions disconnected from your body. Many of your thoughts and
emotions are intimately entwined with how your body interacts with its
environment, so any preserved connectome, to be fully operational as
re-creating the experience of what it is like to be a sentient being, would
also need to be housed in a body. So we would need a warehouse of
brainless clones or very sophisticated robots prepared to have these
uploaded mind neural units installed. How many? Well, to avoid the
charge of elitism, it's only fair that everyone who ever lived be resurrected, so that means multiplying the staggering data package for one
person by 108 billion.

Then there's the relationship between memory and life history. Our
memory is not like a videotape that can be played back on the viewing
screen of our minds. When an event happens to us, a selective impression
of it is made on the brain through the senses. As that sense impression
wends its way through different neural networks, where it ends up
depends on what type of memory it is. As a memory is processed and
prepared for long-term storage we rehearse it, and in the process it is
changed. This editing process depends on previous memories, subsequent
events and memories, and emotions. This process recurs trillions of
times in the course of a lifetime, to the point where we have to wonder whether we have memories of actual events, or memories of the
memories of those events, or even memories of memories of memories . . . What's the "true" memory? There is no such thing. Our memories are the product of trillions of synaptic neuronal connections

that are constantly being edited, redacted, reinforced, and extinguished, such that a resurrection of a human with memories intact will depend on *when* in the individual's life history the replication or resurrection is implemented.

Most of our memories are lost over time, so when God, Omega, the Singularity, or far future Humans (GOSH) reconstruct the pattern of your memories, which ones actually represent you? The answer is *none*, *some*, and *all*. There is no coherently fixed individual in some absolute sense. Our self—our *soul*—consists of a constantly changing matrix of traits and memory patterns that is coherent enough for us to feel that we have a self/soul, and for others to treat us as if we do, so a replicating entity must determine which set of patterns best represents our self/soul such that it would be recognizable to yourself and others. If GOSH resurrects you, for example, which of your memories will be included, and from which point in your life? If it is a select set of memories at some point, say age twenty-nine, that's not all of you. If it is all of the memories you formed throughout your entire life, that might be interesting (and revealing!) but this would not be what it is like to *be* you at any point in your life.

Then there is the continuity of personal perspective problem. If GOSH creates a duplicate Michael Shermer in the far future, how is that any different from the identity *Gedankenexperiment* with the two Michael Shermers in the near future we encountered earlier? When I close my eyes the final time here on Earth, will I open them and be looking out at the far future of the universe in a manner similar to the way I wake up each morning? I don't think so.

Finally, there is the problem of history and the lost past. I have defined history as "a conjuncture of events compelling a certain course of action by constraining prior events."[17] Most of those constraining prior events—contingencies and necessities, or chance and law—are not only lost to historians; they aren't even apparent to those alive at the time. The problem of the irrecoverable past of both people and society is a serious one that any theory of immortality must solve. Even if GOSH could create a perfect replica of someone's genome and connectome, a human life is so much more than that. It is a product of all our relations with other people and *their* life histories, plus our interactions with all the elements in our environment, which is itself a product of countless systems and histories all wrapped up in a complex matrix with so many variables that it is inconceivable how any supercomputer or omnipotent deity could duplicate it all even if the information were available, which it isn't.

———

In his book *The Physics of Immortality,* the physicist Frank Tipler calculates that an Omega Point computer in the far future will contain 10 to the power of 10 to the power of 123 bits (a 1 followed by 10^{123} zeros), powerful enough, he says, to resurrect everyone who ever lived.[18] That may be—it is a staggeringly large number—but is even an Omega Point computer powerful enough to reconstruct all of the historical contingencies and necessities in which a person lived, such as the weather, climate, geography, economic cycles, recessions and depressions, social trends, religious movements, wars, political revolutions, paradigm shifts, ideological revolutions, and the like, on top of duplicating our genome and connectome? It seems unlikely, but if so, GOSH would also need to duplicate all the individual conjunctures and interactions between that person and all other persons as they intersect with and influence one another in each of those lifetimes. Then multiply all that by the 108 billion people who ever lived or are currently living. Whatever the number, it would have to be even larger than the famed Googolplex (10 to the power of a googol, with a googol being 10^{100}, or $10\wedge10^{100}$) from which Google and its Googleplex headquarters derived their names.[19] Even a googol of googolplexes would not suffice. In essence, it would require the resurrection of the entire universe and its many billions of years of history. Inconceivable.

CHAPTER 8

AFTERLIFE FOR ATHEISTS

Can Science Defeat Death?

People say they don't want to live forever. Often their objection is that they don't want to live hundreds of years the way the quintessential 99-year-old is perceived to be living—frail or ill and on life support. First of all, that's not what we're talking about. We're talking about remaining healthy and young, actually reversing aging and being an ideal form of yourself for a long time. They also don't see how many incredible things they would witness over time—the changes, the innovations. Me, I'd like to stick around.

—Ray Kurzweil, *Transcendent Man*, 2009[1]

In the second half of the twentieth century there arose a number of groups and movements dedicated to extending the human life span into centuries, millennia, or beyond—possibly even forever. It is a colorful cast of characters, many of whom I have met and come to know well enough to assure readers that this is no cult movement or financial scam exploiting the fearful. They are the *cryonicists, extropians, transhumanists, Omega Point theorists, singularitarians,* and *mind uploaders,* and they are serious about defeating death.

FREEZE—WAIT—REANIMATE: THE CRYONICISTS

My interest in cryonics is personal, with a threefold connection. It begins with the first human ever frozen in cryonic suspension—James Bedford, a Glendale College psychology professor who, on January 12, 1967, was frozen in liquid nitrogen at -321° F after succumbing to cancer. Having taught at Glendale College for eleven years, I heard anecdotes about "Ol' Doc Bedford" and his eccentric manners (he once taught a course in vocational preparation in which he advised his students, among other things, to wash their underwear at least once a week). It was, evidently, eminently logical that he be first.

Second, for many years my sister Shawn Shermer worked in a research laboratory near Davis, California, for scientists doing cryonics experiments on animals. Sponsored by the American Cryonics Society, the laboratory successfully "chilled" both dogs and monkeys for several hours, then brought them back to active life. The appearance of the beagles Miles and Misty on the popular Phil Donahue talk show brought national attention to cryonics. What was not made clear, however, was that these animals had their temperatures *lowered*; they were not actually frozen in liquid nitrogen as Bedford and dozens more have been.

Third, when we founded *Skeptic* magazine and the Skeptics Society with our monthly science lecture series at Caltech, it seemed natural to take a skeptical look at cryonics. Mike Darwin from Alcor Life Extension Foundation gave a thoughtful presentation to our members, and we published a skeptical analysis in the second issue of the magazine.

Just as most religions have gods and holy books, cryonics offers something of its own trinity of godhead figures and founding documents in Robert Ettinger and his book *The Prospect of Immortality*, Eric Drexler and his book *Engines of Creation*, and Ralph Merkle and his document *The Molecular Repair of the Brain*. Merkle outlines the vision of cryonics and how it will change our lives:

> Disease, disability, and the infirmities of old age will become rarities; joining polio, the plague, and smallpox as ancient pestilences finally laid to rest by the inexorable advance of technology. While some of us might have the good fortune to stay alive and healthy until these medical marvels are available, cryonics offers a bridge to the future in case we wouldn't otherwise make it. Should our health fail we can be cryopre-

served and cooled to the temperature of liquid nitrogen. At that temperature, tissue remains essentially unchanged for centuries. We can preserve ourselves for the few remaining decades until the day when nanomedicine can heal our injuries and restore our health: as good [as] or better than the health we enjoyed in our 20's or 30's.

Since I wrote about cryonics in my September 2001 column in *Scientific American*, Merkle and I have maintained a correspondence as he has tried to convince me to be more open-minded on the possibilities of cryopreservation. "You're still being cited as a cryonics denier," he wrote me in 2014, citing a newspaper article quoting me. "Turn away from the dark side, there's still time!"[2] He points out, for example, that the new techniques used today involving the cryopreservation and vitrification of the brain—which involves turning the cryopreserved brain into a glasslike substance—are superior to the older techniques used on the early adopters decades ago. Merkle has plied me with articles and links, all toward the end of arguing that one's "self" is stored in memory, and it is memory that is being cryopreserved:

> We have evidence that modern cryopreservation methods do, in fact, provide a quality of cryopreservation that is more than enough to preserve human long term memory in the information theoretic sense. The presence or absence of a synapse, as well as the proteins associated with the pre- and post-synaptic structures, and the proteins present in the synaptic cleft, should all be inferable following cryopreservation by today's methods.[3]

The "information theoretic" definition of death means that you aren't dead until your memories are erased, but it remains to be seen whether memories can be recovered after cryopreservation. It is one thing to see intact synaptic structures in a cryopreserved slice of brain in a microscope; it is quite another to see those synapses working to produce memories in a living brain, which has yet to be done. Still, Merkle's point is that cryonics is possible in principle:

> The human brain is physical, and human long term memory is associated with physical changes whose presence would still be identifiable following cryopreservation. Computational power will increase

enormously in the future, as will our ability to image and analyze the changes that have occurred in the cryopreserved human brain. Given your cryopreserved brain, and sufficient computational power, and a sufficient imaging technology, we will be able to recover the information that defines who you are. We will also be able to restore the cryopreserved human brain to a fully functional state.[4]

Given the fact that no one currently frozen has been brought back to life, this is an assertion, not an observation. The renowned neuroscientist Christof Koch, whom I queried on this matter, agreed that an experimental test would be revealing, but he voiced his skepticism about the vitrification of brains:

> As of today, we have no evidence that a vitrified brain can be turned on again later with all memories coming back. This could be tested in mice, for instance, by inducing a specific memory, e.g., using place-specific aversive condition, and then test for that memory after the vitrification process. And it's silly to claim that vitrification doesn't affect the molecular distribution of the roughly 10^6 proteins (from 10^3 different kinds of proteins) present at a pre-/post-synaptic junction. Indeed, it would be utterly amazing if such a highly invasive process didn't profoundly disturb their distribution.[5]

The fact is that the best evidence we have from cryobiology is that it is extremely unlikely that anyone frozen to date will ever be brought back to life. It is one thing to freeze sperm, eggs, or even embryos and bring them back to life; it is quite another to do the same with large organs like the brain. According to Dr. Mehmet Toner, a Harvard Medical School professor of biomedical engineering and a cryobiologist at Massachusetts General Hospital, even the more advanced method of vitrification in which tissues are infused with cryoprotectants is not as effective as cryonicists make it out to be. As he explained in a film for the HBO series *Vice* (which I also participated in), aptly titled *Frozen Faith*, "you can only *slowly* freeze and *slowly* warm large things. Ice is going to form inevitably during warming." In a brain, this would mean neurons and their synaptic connections would be shattered, along with any memories they might retain. Although as we shall see, if the concentration of a cryoprotectant agent is sufficiently high, then rewarming may not cause ice to form, but the evidence for this conclusion is based on tissues much simpler than brains, and on animal brains much smaller than humans'.

Furthermore, current cryoprotectants work on only one type of cell at a time, requiring different agents for different cells. But the human body and brain consists of many different cell types, so protecting one for freezing would mean the sacrifice of others. For example, you might be able to protect many of the cells of a kidney, but you could not also do so of a different organ at the same time, such as a brain. The *Vice* film demonstrated the problem by showing one cell line relatively intact while another cell line treated simultaneously by the same cryoprotective agent was entirely compromised. "This is a very complex problem," Toner continued. "They make it sound like if you vitrify everything will survive. That's not true. So the chances of getting that head back with intact memories is a ridiculous concept." As he reflected, "I've spent thirty-three years day in and day out thinking about how to freeze things, and I know it's not going to work."[6]

This means that the burden of proof is on cryonicists to show that it works, not on scientists to disprove it (or to prove it is unlikely). Here is what would do it for me: cryopreserve and freeze a large mammal such as a dog at -130° Celsius for a week and bring her back to life with her memories relatively intact—she recognizes her name, owner, and home, remembers learned tricks like how to sit up or fetch a stick or ball, and generally acts like the same dog as she was before as judged by her family. *That* would be a proof of concept about which even the most ardent skeptic would have to take notice.

AGAINST THE SECOND LAW OF THERMODYNAMICS: THE EXTROPIANS

As the name suggests, extropians are against entropy. Given the formidable power of the Second Law of Thermodynamics, which holds that the universe is in a state of entropy, these are bold thinkers indeed. For starters, extropians engage in lifestyles meant to delay the inescapable decay of the flesh long enough for the anticipated technological miracles that will transform their bodies into eternal carriers of their patterns of information that constitute a scientific soul. According to the Wikipedia entry for "extropy," the term was coined in 1988 and signifies "the extent of a living or organizational system's intelligence, functional order, vitality, energy, life, experience, and capacity and drive for improvement and growth.[citation needed]" No citation is needed to feel the techno-optimism that exudes from this movement, starting with the noms de plume of the founders Tom Morrow (Tom Bell), Max More (Max T. O'Connor), and Natasha Vita-More (Nancie Clark), the latter said to be the first female transhumanist philosopher (she hopes not to be the last). In his founding document, *The Principles of Extropy*, Max More (who is now the CEO of the cryonics organization Alcor) outlines the principles of extropy that resonate with reasonableness. First, extropians believe in perpetual progress, which "means seeking more intelligence, wisdom, and effectiveness, an open-ended lifespan" on the positive side of the ledger, and "the removal of political, cultural, biological, and psychological limits to continuing development" on the negative side. Second, extropians affirm "continual ethical, intellectual, and physical self-improvement, through critical and creative thinking, perpetual learning, personal responsibility, proactivity, and experimentation." Third, extropians take action "with positive expectations" with both individuals and organizations being "tirelessly proactive." Fourth, extropians design and manage "technologies not as ends in themselves but as effective means for improving life." And fifth, extropians support "social orders that foster freedom of communication, freedom of action, experimentation, innovation, questioning, and learning."[7]

Well, who am I to object to such lofty and liberating goals? But it is one thing to work toward progress in incremental steps in hopes of achieving modest and realizable goals; it is quite another to accomplish the far loftier aims of the extropians as More envisions them: "We have achieved two of the three alchemists' dreams: We have transmuted the elements and learned to fly. Immortality is next." Citation needed.

Transmuting elements and learning to fly are spectacular achievements, but they're modest protopian steps compared to the Grand Canyonesque leap of unimaginable distance required to conquer death. Our mortality appears to be programmed into every cell, organ, and system in our bodies such that immortality will require the solving of numerous problems at many levels of complexity, all at the same time. Even if we manage to break through the upper ceiling of approximately 125 years by solving these many problems, who knows what additional medical issues may arise that we cannot as yet conceive if we lived, say, two hundred, five hundred, or a thousand years. Instead of reaching for the utopian goal of immortality, or a thousand years, a more modest goal of living to age 150 at a relatively high quality of living (and not in a nursing home bed hooked up to a feeding tube and breathing machine) would be something well worth aiming at.

FROM HUMAN TO POSTHUMAN: THE TRANSHUMANISTS

Closely related to the extropians are the *transhumanists* (or H+, as they're known), who intend to transform the human condition first through lifestyle choices involving diet and exercise, then through body enhancements (e.g., breast or cochlear implants) and body parts replacements (e.g., artificial knees, hips, hearts, or livers). H+ers propose that such steps will lead to a newfound "morphological freedom" that will be further advanced through genetic engineering, all with the goal of taking control of evolution and transforming the species into something stronger, faster, sexier, healthier, and with vastly superior cognitive abilities the likes of which we mere mortals cannot conceive.

The transhumanist movement goes far beyond the secular humanist movement that arose in the twentieth century as a replacement for religion, because "humanism tends to rely exclusively on educational and cultural refinement to improve human nature," More writes in his introduction to *The Transhumanist Reader* (the definitive reference for all things H+), "whereas transhumanists want to apply technology to overcome limits imposed by our biological and genetic heritage." More's reach certainly exceeds his grasp:

> Transhumanists regard human nature not as an end in itself, not as perfect, and not as having any claim on our allegiance. Rather, it is just one point along an evolutionary pathway and we can learn to reshape our own nature in ways we deem desirable and valuable. By thoughtfully, carefully, and yet boldly applying technology to ourselves, we can become something no longer accurately described as human—we can become posthuman.[8]

Just as we are already enhancing our senses with glasses, hearing aids, cochlear implants, and wearable computerized clothing, there is no reason why we could not also enhance our brains. We are already doing this with paralyzed patients who undergo surgery to have a computer chip implanted in the motor cortex of their brains and then learn to control an artificial arm or a computer cursor to read and write, all just by thinking. The computational neuroscientist (and transhumanist) Anders Sandberg has computed the possibility that such enhanced computer brains could be scaled up to "Jupiter-sized brains" whose intelligence would so far surpass us that we would be like lab rats to them.[9] The philosopher Mark Walker concludes that it would not be hyperbole to suggest "those who upload may well be on their way to godhood."[10]

One of the more intriguing transhumanists I met was a man named Fereidoun M. Esfandiary, FM-2030 for short (the date of his hundredth birthday and that of the hoped-for singularity), who reminds me of the Most Interesting Man in the World character of the Dos Equis beer commercials ("His passport requires no photograph"; "He can speak Russian . . . in French"). When I could not place his accent (he is the son of an Iranian diplomat and had lived in seventeen different countries by age eleven), he told me that the world is his country and that he rejects traditional collectivist notions such as nationality. A handsome man who looked ageless, he proclaimed when I queried him about that, "I have no age. Am born and reborn every day. I intend to live forever. Barring an accident I probably will." If you can make it to 2010, he told Larry King in a 1990 interview, you will probably make it to 2030, and "if you are around in 2030 there's an excellent chance you can coast to immortality."[11] Unfortunately, FM-2030 didn't make it to 2010, much less 2030. He was struck down by pancreatic cancer in 2000 and now resides in a vat of liquid nitrogen at the Alcor Life Extension Foundation in Scottsdale, Arizona. This line from the Most Interesting Man in the World commercials is fitting: "Time waits on no one . . . but him." We shall see. For now, FM-2030 is a posthuman, not a transhuman.

THE ALPHA AND OMEGA OF IMMORTALITY: OMEGA POINT THEORISTS

Omega Point Theory (OPT) contends that one day we will all be resurrected in a super-powerful virtual reality that is so authentic in detail that it is indistinguishable from physical reality today. This is different from the theory that contends that our entire universe is a *Matrix*-like virtual reality in some extraterrestrial's computer, an idea supported by the estimable philosopher Nick Bostrom.[12] What OPT suggests, however, is not so different in the sense that Bostrom thinks the computer simulation is running now, whereas OPT places it in the far future.

In OPT there is no need to fear death, or have yourself cryonically frozen, or have your memories uploaded into a computer, because in the far future of the universe, computers will be so powerful that they will be able to re-create every human who ever lived. OPT's most prominent champion is Frank J. Tipler, a physicist and Christian who believes his theory of physics perfectly matches the biblical narrative of the cosmos and humanity, which he defends in two books, *The Physics of Immortality* and *The Physics of Christianity*.[13] I have written about Tipler and his ideas in detail elsewhere,[14] and I have gotten to know him well enough to say with confidence that he absolutely believes what he says about the inevitability of immortality, and that this outcome was foretold in the Bible, albeit in language appropriate for the time.

Like cryonicists, extropians, and transhumanists, Tipler is a techno-optimist who believes in humanity's cosmic destiny, much like his childhood hero, the German rocket builder (V-2, Saturn V) Wernher von Braun. "The attitude of unlimited technological progress is what drove Wernher von Braun and it is what has motivated me all my life," Tipler told me in an interview.[15] Unlike most cryonicists, extropians, and transhumanists, who tend to be atheists or agnostics, however, Tipler is a thoroughgoing Christian who accepts all the major tenets of that religion as if they were derivatives of physics equations. Take the soul, for instance. "A soul is either a pattern in matter or a mysterious soul substance," Tipler explained, noting that "Plato took the position that the soul consists of this soul substance, whereas Thomas Aquinas took the attitude that resurrection was going to be reproducing the pattern, which is what I argue in my book." How is a soul pattern reproduced?

Tipler builds on the Jesuit priest Pierre Teilhard de Chardin's idea of the Omega Point, a conviction that the universe is inexorably evolving toward a higher plane of consciousness, at the end of which evolution there is a unity of creation between humans and God. Teilhard had a vague notion of technological progress helping to bring this about, which Tipler—and, as we shall see below, *singularitarians*—have filled in with scientific proposals of how, precisely, this could happen. "The instant the Omega Point is reached, life will have gained control of all matter," Tipler explained in his book *The Anthropic Cosmological Principle*. By that time, "life will have spread into all spatial regions in all universes which could logically exist, and will have stored an infinite amount of information, including all bits of knowledge which it is logically possible to know. And this is the end."[16] In physics and cosmology, a point of infinite energy, density, and information is called a singularity, said to be the Big Bang starting point of the universe and the Big Crunch ending point—the Alpha and the Omega. The singularity in physics corresponds to eternity in religion, says Tipler. When I asked him to summarize his theory in a single sentence, he composed: "Rationality increases without limit; progress goes on forever; life never dies out."

At the end of time, Tipler predicts, the universe will collapse, providing enough energy from the collapsing process for supercomputers to re-create every human who ever lived in a virtual reality indistinguishable from our reality. Since this far future supercomputer is, for all intents and purposes, omniscient and omnipotent, it is God. This deity, Tipler presumes, will want to re-create us all in its virtual reality. That is the resurrection—immortality in a souped-up version of the *Star Trek* Holodeck, a virtual reality so authentic that you couldn't tell the difference from our current reality, nor could you turn it off by commanding, in *Star Trek* lore, "Computer, end program."

I have already outlined six reasons why I am skeptical of Omega Point Theory in my book *Why People Believe Weird Things*[17] so I won't belabor my skepticism unnecessarily here except to say, in brief, that OPT is purely theoretical without an empirical basis, it depends on science and technology continuing unabated from the past into the future at accelerating rates (which may not happen), and it has far too many "if-then" propositions, all of which have to be true (such that if any one fails then the theory unravels). Moreover, OPT too conveniently matches our verdant hope to live forever and reconnect with our lost loved ones, and too perfectly matches the religion of its chief advocate. I am always skeptical of scientists who concoct theories to explain all the events and miracles of the Bible. I would be more impressed if a scientist discovered that, say, 46 percent of the stories in the Bible were true and the other 54 percent were myth, because this would indicate more objective integrity and less motivated reasoning.

In *The Physics of Christianity*, for example, Tipler suggests that the Star of Bethlehem was a supernova erupting in the Andromeda galaxy, timed to signal the birth of Jesus. The virgin birth is explained by parthenogenesis, or reproduction without sex, which Tipler thinks can be proven by testing the blood on the Shroud of Turin, which he accepts as genuine even though it is carbon-dated to the fourteenth century. Tipler thinks Jesus walked on water by "directing a neutrino beam" downward from his feet into the water, and that he ascended into the clouds to heaven using the same technology of beamed neutrinos. Finally, the physics of Jesus' resurrection involved the atoms in his body spontaneously decaying into neutrinos and antineutrinos "in a fraction of a second, after which the energy transferred to this world would have been transferred back to the other worlds from whence it came." Here Tipler refers to the "many worlds interpretation" of quantum mechanics in which there exists an infinity of universes, some of which are similar to our own and include exact duplicates of all of us.

Maybe none of this needs explaining because Bible stories present myths with moral homilies instead of historical facts. But that's a problem that takes us too far afield. Since this is a theory of physics, I turn to an analysis of Tipler's theory by the physicist Lawrence Krauss, who points out that Tipler's claim that the standard model of particle physics is complete and exact is not true. Further, says Krauss, Tipler's claim that we have a clear and consistent theory of quantum gravity is also untrue. Tipler says that the universe must recollapse for his theory to be true. Krauss notes that "all evidence thus far suggests that it won't." Finally, Krauss concludes about Tipler, "He argues that we understand the nature of dark energy. We don't. He argues that we know why there is more matter than antimatter in the universe. We don't. I could go on, but you get the point."[18] The point is that the Omega Point's promise of immortality fails to deliver on its scientific promise.

TRANSCENDENT MAN: THE SINGULARITARIANS

As the name suggests, singularitarians are scientists considering singularity-level technologies to engineer immortality by, among other things, transferring your soul—the pattern of information that represents your thoughts and memories as stored in the connectome of your brain—into a computer. In the 2014 film *Transcendence*, Johnny Depp's scientist character, Dr. Will Caster, has been working on a computer capable of achieving sentience when he is attacked by terrorists and shot with a polonium-laced bullet that leaves him with a month to live. Just before he dies, Caster uploads his mind into his quantum computer such that the continuity of his personal perspective is not broken. His self has moved from a biological medium to a silicon platform. Since there was no distinct break between conscious states, Dr. Caster's continuity of self continues. He has transcended his body. He is his pattern, not his matter. He then wants to go online because, as he explains to an audience of neuroscientists, mathematicians, and hackers, "once online, a sentient machine will quickly overcome the limits of biology. And in a short time its analytic power will become greater than the collective intelligence of every person born in the history of the world. So imagine such an entity with a full range of human emotion. Even self-awareness. Some scientists refer to this as 'the singularity.' I call it 'transcendence.'"

Art imitates life. Johnny Depp's real-life singularity scientist is Ray Kurzweil, aka the "transcendent man" in Barry Ptolemy's documentary film of that title.[19] "Transcendent" is the right adjective for this scientist, futurist, author, and inventor of such life-changing technologies as the first optical character recognition program and CCD flatbed scanner, the first print-to-speech reading machine for the blind, the first text-to-speech synthesizer, Kurzweil electronic keyboards, and more. At age fifteen he was designing computer programs to aid in homework, and at age seventeen he won the Westinghouse Science Talent Search contest, which landed him an invitation to the White House. Recipient of the 1999 National Medal of Technology and inductee into the National Inventor Hall of Fame, Kurzweil wrote *The Age of Intelligent Machines* and *The Age of Spiritual Machines*, which influenced the field of artificial intelligence, and his book *The Singularity Is Near* popularized the term and the hope that one day soon we will live forever.

Kurzweil's motivation stems in part from the premature death of his father at age fifty-eight. Fredric Kurzweil was a professional musician who, Ray's mother explains, was never around while his boy was growing up. Like father, like son—Kurzweil's own workaholic tendencies in his creation of over a dozen companies starting at age seventeen meant he never really knew his father. As the film *Transcendence* portrays the tormented inventor, Kurzweil's mission in life seems more focused on resurrecting his patriarch than in resuscitating humanity. An especially lachrymose moment is when Kurzweil is riffling through his father's journals and documents in a storage room dedicated to preserving his memory until the day that all this "data" (including Ray's own fading memories) can be reconfigured into an AI simulacrum so real that the son will feel reunited with the father he never knew. Through heavy sighs and wistful looks, Kurzweil appears less a proselytizer on a mission than a man tormented. In one scene Kurzweil is shown wiping away a tear at his father's grave site; in another he pauses over photographs and looks longingly at mementos. Although Kurzweil says he is optimistic and cheery about life, he can't seem to stop talking about death: "It's such a profoundly sad, lonely feeling that I really can't bear it," he admits. "So I go back to thinking about how I'm not going to die."

How does Kurzweil plan not to die? It begins with what he calls "the law of accelerating returns," which holds not just that change is accelerating, but that the *rate* of change is accelerating. Moore's Law has accurately projected the doubling rate of computer power since the 1960s. The singularity is Moore's Law on steroids and applied to all science and technology. Before the singularity, the world will have changed more in a century than it had in the previous thousand centuries, which is staggering enough. As we approach the singularity, says Kurzweil, the world will change more in a decade than in a thousand centuries, and as the acceleration continues and we reach the singularity, the world will change more in a year than in all presingularity history. When that happens, humans will achieve immortality.

Postsingularitarians will be to us what we are to our pets: so vastly smarter that we won't even know how intelligent they are. Within a quarter century, Kurzweil projects, "nonbiological intelligence will match the range and subtlety of human intelligence," then "soar past it because of the continuing acceleration of information-based technologies, as well as the ability of machines to instantly share their knowledge."[20] Compare the room-sized computers of the 1950s to the pocket-sized computers we carry around in our pockets and then follow that trajectory downward in size over the same amount of time, or less, and you arrive at cell-sized computers that can be digested in tablet form. Once such nanotechnologies exist in the form of nanorobots that will repair cells, tissues, and organs (including brains), when they are coupled with other biotechnologies like designer drugs and engineered genes, the aging process will be halted, and possibly even reversed, enabling us "to live long enough to live forever," as he proclaimed in his book *Transcend*.[21]

To secure your health until this secular second coming (around 2040), Kurzweil's book, *Fantastic Voyage: Live Long Enough to Live Forever* (coauthored with Terry Grossman), recommends that we adopt "Ray and Terry's Longevity Program," which includes 250 supplements a day and weekly rounds of biochemistry reprogramming through intravenous nutritionals and blood cleansing. To boost antioxidant levels, for example, Kurzweil suggests a concoction of "alpha lipoic acid, coenzyme Q_{10}, grapeseed extract, resveratrol, bilberry extract, lycopene, silymarin, conjugated linoleic acid, lecithin, evening primrose oil (omega-6 essential fatty acids), n-acetyl-cysteine, ginger, garlic, 1-carnitine, pyridoxal-5-phosphate, and Echinacea."[22] Bon appétit.

The singularity as envisioned by Ray Kurzweil is admittedly inspiring. He is not a big man, but onstage in full Singularity Is Near mode, he is larger than life. Here he is in 2016, with the full backing of the tech giant Google behind him as their director of engineering, explaining in a *Playboy* interview what we have to look forward to:

> By the 2030s we will have nanobots that can go into a brain non-invasively through the capillaries, connect to our neocortex and basically connect it to a synthetic neocortex that works the same way in the cloud. So we'll have an additional neocortex, just like we developed an additional neocortex 2 million years ago, and we'll use it just as we used the frontal cortex: to add additional levels of abstraction.

Not just smarter, but healthier:

> As they gain traction in the 2030s, nanobots in the bloodstream will destroy pathogens, remove debris, rid our bodies of clots, clogs and tumors, correct DNA errors and actually reverse the aging process. I believe we will reach a point around 2029 when medical technologies will add one additional year every year to your life expectancy.[23]

As the rate of progress of medical technology accelerates, the years will pile up for decades, centuries, and beyond, possibly to forever.

Kurzweil is no armchair philosopher. He works for Google, and his bosses, Larry Page and Sergey Brin, started the biotech company Calico to develop the science and technology to expand the human life span to over two hundred years. Hedge fund manager and PayPal cofounder Peter Thiel created Breakout Labs in order to fund scientists and start-ups that are working on achieving immortality, and he invested $3.5 million into the antiaging Methuselah Foundation, founded by Aubrey de Grey, a biomedical gerontologist who treats aging as an engineering problem to be solved at the cellular level by reprogramming the anatomy, physiology, and genetics of cells so that they stop aging. Oracle cofounder Larry Ellison has contributed more than $430 million toward antiaging research because he finds the quiet acquiescence of mortality "incomprehensible." As he told his biographer, "Death has never made any sense to me. How can a person be there and then just vanish, just not be there?"[24] It's a good question, and one that others are working on through mind uploading.

UPLOADING THE MIND AND PRESERVING THE CONNECTOME

Your connectome is the pattern of information—your thoughts and memories—that represents your MEMself and, as long as you're alive, your POVself.[25] Research on copying your connectome is something of an extension of cryonics, which also depends on upholding the structural integrity of the brain because without an intact connectome, individuals brought back to life without their thoughts and memories would not be themselves. They would either be unique MEMselves with new memories that formed once they started functioning again, or else they would be zombies. There are now scientists working on preserving the connectome in such a way that it can be stored for centuries, or even millennia, unchanged, until the day comes when it can be uploaded into a computer and "turned on" like Johnny Depp's character in *Transcendence*.

The preservation of the connectome is one of the goals of 21st Century Medicine in Fontana, California.[26] The company specializes in the cryopreservation of human organs and tissues using cryoprotectants (antifreeze) so that they may be excised, transported, and transplanted into new patients with minimum damage. In 2009, for example, the lab's chief research scientist, Gregory M. Fahy, published a paper in the peer-reviewed journal *Organogenesis*, documenting how his team successfully transplanted a rewarmed rabbit kidney after it had been cryo-protected and frozen to -135° C through the process of vitrification, "in which the liquids in a living system are converted into the glassy state at low temperatures."[27] If kidneys can be so preserved, why not brains?

Fahy and his colleague Robert L. McIntyre have developed a technique that they employed to win a portion of the Brain Preservation Prize (BPP), established by the neuroscientist Kenneth Hayworth, a senior scientist at the Howard Hughes Medical Institute and president of the Brain Preservation Foundation. I am on the advisory board of the Brain Preservation Foundation (BPF) as something of an *advocatus diabolic*, so I follow the research closely. When it became apparent that they were in the running for the prize, I was invited to tour the facilities of 21st Century Medicine in September 2015 to meet their principal investigators and observe the research firsthand. The prize is currently valued at $106,000 (private donations continue to increase it), the first 25 percent of which was awarded to the 21st Century Medicine team in February 2016 for the Small Mammal Phase of the prize for the complete preservation of the synaptic structure of a whole rabbit brain. No one has yet won the other 75 percent for the first team "to successfully preserve a whole large animal brain in a manner that could also be adopted for humans," but two teams have already submitted specimens for analysis, and the prospects are good that someone will win the second phase of the prize fairly soon.

After meeting Fahy, McIntyre, and their support staff and fellow scientists, Hayworth and I were escorted to the main laboratory room, where we were shown the Controlled Isothermic Vapor Storage units. These huge cylindrical tanks are filled with liquid nitrogen that released clouds of vaporous gas when we opened the lids to examine the plastic containers containing frozen whole brain samples prepared for the BPP, a total of three rabbit brains and two pig brains. Neither of the rabbit brains showed any visible signs of ice formation or damage, but one of the pig brains had a dime-sized ice smudge in the occipital lobes near the cerebellum. Not a good sign, but this is not as yet a perfect science.

From there we moved to the surgical bay, where there was already a rabbit lying on the table unconscious under anesthesia.[28] The surgery on the rabbit began by shaving the fur off its neck in order to access the carotid arteries, which were carefully opened so that a tiny plastic tube could be inserted into each, through which the brain was infused with a fixative chemical called glutaraldehyde that acts as a preservative. When injected into a brain, within minutes it stabilizes neural structure by binding the proteins within the neurons together into a solid gel. Brains fixed in this manner can retain structural stability at room temperature for several days with no degradation. At this point in the process, when I wondered aloud about what all this was doing to the rabbit, I was told that the animal was dead and could not be brought back to life. Oh.

After about forty-five minutes of preservative perfusion, a cryoprotectant agent was pumped into the rabbit's brain that then allowed it to be lowered to the storage temperature of -135° C, a temperature so low that the brain vitrifies without ice crystal formation. The next day Hayworth processed the brain, utilizing a machine that cuts it into razor-thin slices of 150 μm each (a μm is a micrometer, or micron, that equals one millionth of a meter, and 150 μm is about the thickness of a single human hair). The slices were then put on microscopic slides and viewed and analyzed through an electron microscope to determine if there was any damage from the fixation, vitrification, or freezing processes. The result was an intact glutaraldehyde fixed brain with no visible macroscopic defects; that is, there was no ice damage and the structure of the neurons and their synapses was intact.[29] This entire process is called *aldehyde-stabilized cryopreservation* (ASC), and it was in fact the slices from the rabbit brain that I witnessed being prepared that won the Brain Preservation Prize. Figure 8-1 shows the liquid nitrogen containers that hold the frozen brains and the temperature gauge showing that they are stored at -125° C, the surgical bay at 21st Century Medicine where the surgery on the rabbit was conducted, and the microscope through which I observed the brain slices to assess the structural integrity of the neurons and their synapses after preservation.

The experiment, McIntyre told me, was a proof of concept.[30] How is a dead animal proof of anything? I wondered aloud. Think of a book in epoxy resin hardened into a solid block of plastic, McIntyre parried. "You're never going to open the book again, but if you can prove that the epoxy doesn't dissolve the ink the book is written with, you can demonstrate that all the words in the book must still be there ... and you might be able to carefully slice it apart, scan in all the pages, and print/bind a new book with the same words." In this analogy, McIntyre

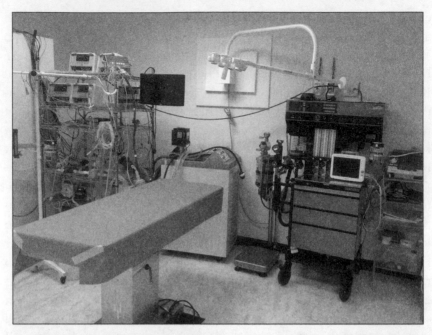

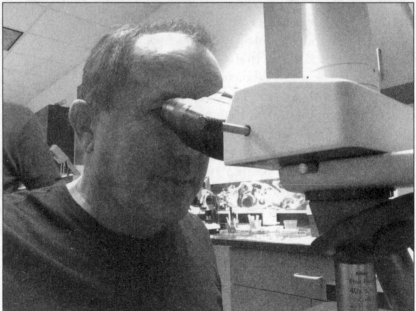

Figure 8-1

The 21st Century Medicine surgical bay where I observed the surgery on the rabbit brain, and the microscope through which I observed the rabbit brain slices to assess its structural integrity after preservation. From the author's collection.

continued, "brain viability is like the ability for you to open a book, leaf through the pages, and read the book's story. The brain's connectome is like the words on the pages of the book. You can preserve the words while making the book impossible to open, and you can prove that the words are preserved by doing your preservation procedure on test books, cutting them open, and seeing that the words haven't been altered." This, McIntyre explained, "is what I'm trying to do with brains—the glutaraldehyde glues all the proteins together while maintaining the structural connections between all the brain's cells. Storage at -135° C completely locks everything in place and guarantees long-term storage."

Further examination by Hayworth and the other Brain Preservation Prize judge, the MIT neuroscientist Sebastian Seung, determined that the connectome conservation was such that the prize criteria had been met. Through the 3-D scanning electron microscope, Hayworth told me, the rabbit brain circuitry he examined "looks well preserved [and] undamaged, and it is easy to trace the synaptic connections between the neurons." Thus they have moved on to the large mammal phase of the contest where, Hayworth elaborated, the 21st Century Medicine team "has already submitted to the BPF such a preserved pig brain for official evaluation." Hayworth added that 21st Century Medicine "has already presented credible electron microscopic evidence in their *Cryobiology* publication that their Aldehyde-Stabilized Cryopreservation technique can preserve a whole pig brain, but of course this will have to be verified independently by the BPF to win the final phase of our prize."[31] As of this writing, the BPP for large mammals has yet to be granted.[32] Figure 8-2 shows what the rabbit and pig brains look like before analysis and after.

During my tour of the lab, Ken Hayworth offered a full disclosure to me that he has more than a detached scientific interest in this subject. He would like humanity to transcend mortality and is motivated to be a part of that effort, even if it doesn't happen in his lifetime.[33] Assuming the connectome remains intact after cryopreservation, I pressed him, how do you know memories are preserved if, say, such a brain were reawakened? "The most direct evidence that all human long-term memory is stored statically might be the Profound Hypothermia and Circulatory Arrest (PHCA) surgical procedure used since the 1960s. Patients' brains (and cores) are cooled to as low as 10° C during surgeries lasting

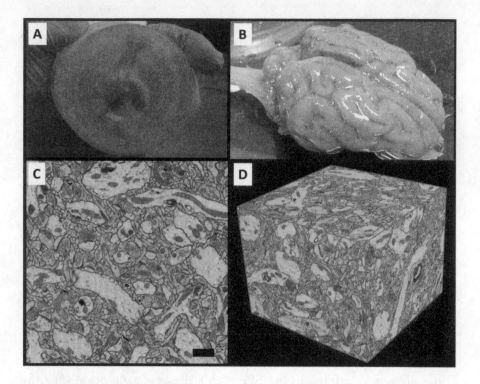

Figure 8-2. Cryopreserved Brains

(A) ASC-preserved rabbit brain vitrified at -135° C. (B) ASC-preserved pig brain rewarmed after being stored at -135° C. The goo covering it is melted CPA. (C) Electron micrograph of cortex sample taken from the rabbit brain shown in (A). Synaptic details are clearly visible and well preserved. Scale bar is 1 micron. (D) 10 × 10 × 8 micron FIB-SEM volume acquired at 8nm resolution taken from the rabbit brain shown in (A). Neural processes and connections are easily traced in this volume (similar FIB-SEM videos are available online as part of the ASC publication). Courtesy of Kenneth Hayworth.

around thirty minutes.[34] It is known that at temperatures below 20° C, excitatory synapses are no longer able to function. Thus all patterned electrical activity is routinely stopped during a PHCA surgery, but the patients recover with memories and personality intact. This is strong evidence that long-term memories (past a few hours) are stored in a static manner."[35]

A review paper by the Nobel laureate and memory pioneer Eric Kandel and his colleagues summarizing the research on memory through 2015, for example, argues that memories longer than a few hours old are stored as static structural changes that can be seen in electron microscope images.[36] Another 2015 review paper titled "Memory Engram Storage and Retrieval" by the neuroscientist Susumu Tonegawa and his colleagues summarized over a century's worth of experimental research demonstrating, in their technical jargon, that

> memory may be stored in a specific pattern of connectivity between engram cell ensembles distributed in multiple brain regions and this connectivity pattern is established during encoding and retained during consolidation. Based on these integrative findings, we propose that enhanced engram cell-specific synaptic strength is crucial for the retrievability of particular memory engrams, while the memory information content itself is encoded in a pattern of engram cell ensemble connectivity.[37]

A more concise and poetic way to say all this is *neurons wire together if they fire together*.[38] The wires are where the memories are, and they can be changed with changing environments, also known as *neural plasticity*. Thus, underlying the overall connectome for a brain is an engram circuit for a given memory, or a memory *engrome*, as it is called. Each of these engromes constitutes a memory or memory pattern, the totality of which constitutes the connectome. In the words of Eric Kandel in his foundational textbook *Principles of Neural Science*: "One of the chief ideas we shall develop in this book is that the specificity of the synaptic connections established during development underlie perception, action, emotion, and learning."[39] As Hayworth summed it up for me: "In the neuroscience literature there is a truly vast array of evidence that long-term memories are stored as static connections. The 'static system of connections = memories' is as fundamental to current neuroscience theories as 'static sequences of DNA base pairs = genes' is to theories of biology."

THE SCIENCE OF THE SOUL

The analogy between memories and genes is apt because the codiscoverer of the structure of DNA, Francis Crick, famously took a career detour from genetics into neuroscience to uncover the nature of consciousness, and opened his bestselling book *The Astonishing Hypothesis* with these now oft-quoted lines:

> The Astonishing Hypothesis is that "you," your joys and your sorrows, your memories and your ambitions, your sense of personal identity and free will, are in fact no more than the behavior of a vast assembly of nerve cells and their associated molecules. This hypothesis is so alien to the ideas of most people alive today that it can truly be called astonishing.[40]

Crick subtitled his book *The Scientific Search for the Soul*, which I note here because this is the undercurrent driving much of this research; and not just the search for the soul, but the preservation and resurrection of it. Why? I put the question to Hayworth, whose mannerisms and affectations remind me of Sheldon Cooper on *Big Bang Theory* (without the nerdy laugh). "Well, it would be great to see the future firsthand," he began. "As an atheist I know there is nothing to fear in death, the party will just go on without me. But I would very much like to stay at the party a while longer, especially to see us colonize space and unlock the mysteries of the universe." That's hardly unique, as most of us would like to stay at the party for as long as we can, but there is more for Hayworth. Degenerative brain disorders such as dementia and Alzheimer's means that even if the extropians and the transhumanists succeed in extending life by decades or more, unless we can deal with these brain diseases it won't be worth living longer. "When the human race finally achieves the technological and scientific prowess necessary to upload our minds into custom designed mechanical bodies and computerized brains," Hayworth suggested, "those created bodies will have none of the limitations of our biological bodies designed by uncaring natural selection."

That sounds utopian, and Hayworth is realistic enough to realize that if "even a hard-core skeptic like you instinctively rejects the idea that mind uploading is possible," how much harder it is going to be to convince masses of religious people who "look forward to Jesus coming back to destroy the earth and take our souls up to heaven." And that's here in the West. Imagine what radical Islamic terrorists would do if they came upon a facility for brain uploading or a cryonics company, given their propensity to destroy even religious icons of faiths different from their own. Even most of his neuroscience colleagues think Hayworth has gone around the bend on this project, but as he reflected, "Whether I personally get to see the future is inconsequential. What is important to me is that this technology (brain preservation) has the possibility of making the world a much better place. That is worth working toward."

Indeed it is, but will it do what he thinks it will do? That is, can it solve the continuity of personal perspective problem such that when the connectome with all your memories is brought back to life it will really be "you" waking up inside the computer, in a manner similar to you waking up from a long sleep? That is, is the MEMself the same as the POVself? I don't think it is, but Hayworth does, and in support he references Derek Parfit's book *Reasons and Persons*, in which the late philosopher proposes this thought experiment to counter the continuity problem: One person with a split brain whose two hemispheres are identical in every way (including identical memories) sacrifices himself for his two identical brothers who are dying of irreversible brain damage by donating each of his two brain hemispheres to them. Now you have two identical people with the same set of memories walking about in two separate bodies, each feeling psychologically continuous from before when there was just one.[41] What happened to "you"? Nothing. Your old body is gone, but there are now two of you in two separate but identical bodies. Parfit believes that the psychological continuity in splitting and transplanting the two identical brain halves solves the personal perspective problem, as there are now two continuous personal perspectives, each of whom feels like and perceives himself to be "you" in every way. In this sense, then, uniqueness is not the key to identity. It doesn't matter how many there are of you—one, two, lots; as long as there is psychological continuity when the copies are made (or transferred to another platform), then, as in the *Star Trek* episode with the duplicate Riker, you would each begin to lead separate lives, develop new and different memories, and become, once again, unique.

In this view, the self is not "our moment-to-moment ongoing sense of self" or our POVself, Hayworth insists, but "our unique set of memories," or MEMself. He admits that a whole person consists of both POVself and MEMself, but he believes that once the MEMself transfer is made and the computer turned on, the POVself will also be activated. After all, POV is just how you are looking out at the world at any given moment as the information from it streams into your brain through your senses, which all organisms experience no matter how simple or complex. Your dog has a POV, as does the ant crawling across your floor. Every living thing has a POV. The key to selfhood, says Hayworth, lies in the thoughts and memories, which are encoded in MEMself.

I disagree with Hayworth and have told him so. In response, he argues that the sense of POVself as the primary one is an illusion, not unlike the illusion that we see a unified field when we look out into the world even though there is a blind spot in the retina where the optic nerve exits the eye, or the larger illusion that we are one unified self even though the brain consists of numerous neural networks working independently to solve different problems and run different systems. We are simply unaware of all that our brains are doing, and that's a good thing, otherwise the world would be a buzzing blurry chaos of activity. If the self is an illusion, then so, too, is the POVself, because what it's like to be you at any given moment (your point of view looking out through your eyes) is not real. What's real is the totality of *engromes*—all the engrams making up your memory that together with your thoughts composes the connectome—the MEMself.[42]

I still don't see how the MEMself alone can be your self (or your soul). If duplicating the MEMself were done without the death of the person, then there would be two MEMselfs, each with its own POVself looking out at the world through their unique eyes. At that moment each would take a different path in life thereby recording different memories based on different experiences. "You" would not suddenly have two POVs. If duplicating the MEMself were done upon the death of the person, there is no known mechanism by which your POVself would be transported from your brain into a computer (or a resurrected body). As I demonstrated in the *Gedankenexperiment* with the two Michael Shermers in the previous chapter, the MEMself is not the same as the POVself; if you copied my connectome and uploaded it into a computer and turned it on, I do not think that it would be like waking up from a long sleep with a continuity of self intact. A POV depends entirely on the continuity of self from one moment to the next, even if that continuity is broken by sleep or anesthesia. Death is a permanent break in continuity, and your personal POV cannot be moved from your brain into some other medium, here or in the hereafter. Of course, I could be wrong, and I would not protest were I to awaken in some paradisiacal state after death with both MEMself and POVself fully functioning.

THE PTOLEMY VS. COPERNICAN PRINCIPLES:
THINKING SKEPTICALLY ABOUT IMMORTALITY

If there is one thing I have learned in a quarter century of professional skepticism, it is this: Beware the prophet who proclaims the end of the world, the apocalypse, doomsday, or judgment day is upon us, or that the second coming, the resurrection, paradise, or the *Greatest Thing to Happen to Humanity Ever* is coming in the prophet's own lifetime. The belief that we can transcend what no one before us has arises from our natural inclination to assume that we are special and that our generation will witness the new dawn. Call it the *Ptolemaic Principle*—the belief, like its namesake's, that we are not only at the center of the universe but are specially created, chosen people living at a unique time in history. People have always embraced the Ptolemaic Principle, but it is gainsaid by the *Copernican Principle*, which, pace its namesake, holds that Earth is not the center of the solar system, the solar system is not the center of our galaxy, our galaxy is not the center of the universe, humans are not specially created apart from all other animals, and we are not living in the most important time in history.

Recall, too, the *Mediocrity Principle*, which states that an item selected at random from a population is most likely to have come from the most numerous type of those items. Reach into a bag containing a thousand Ping-Pong balls, nine hundred of which are white and one hundred black, the chances are pretty good (90 percent, in fact) that you will select a white ball. Pluck someone at random and chances are that that person will represent the population at large. Select any generation of humans from the past and they are likely to resemble those who came just before or after them. Everyone feels special, and every generation believes they are living in special times, but statistically speaking this cannot be true. Thus, the chances that even a science-based prophecy such as those proffered by cryonicists, extropians, transhumanists, singularitarians, and mind uploaders will come true is highly unlikely. Prophets of both religious and secular doom foretell the demise of civilization within their allotted time (and that they will be part of the small surviving enclave while everyone else is left behind). Prognosticators of both religious and secular utopias always include themselves as members of the chosen few, with the paradisiacal state just within reach. Rarely do we hear a scientific futurist or a religious diviner predict that the "big thing" is going to happen in, say, the year 7510. But where's the hope in that? Yes, proponents of cryonics do make such long-range projections, admitting that the technologies needed to resurrect the frozen dead are likely many centuries in the future, but perhaps that's why they have so few takers.

Evaluating these science-based theories of immortality, cryonics seems like a better bet than mind uploading, only because without the continuity of self through one's point of view (the POVself), having my body, brain, and connectome preserved, frozen, stored, defrosted, warmed, and reawakened feels more like waking up from a long sleep than does having a copy of my connectome uploaded into a computer (the MEMself), assuming any of this ever actually worked. And here we face a type of Pascal's wager: if you do nothing and have yourself buried or cremated upon death, there is a zero chance of returning; if you sign up with one of the cryonics organizations, there is at least a greater-than-zero chance of being resurrected. So . . . should you make arrangements to be frozen, just in case? At Alcor, the oldest and most established cryonics organization in the world, the standard financial plan is to take out a life insurance policy with Alcor as the beneficiary to cover the costs ($200,000 for your full body, $80,000 for just your head). So, depending on your age and health, the premium could be between a few hundred and several thousand dollars a year. The cryonicist Ralph Merkle outlines the choice matrix as follows: if you sign up and it works, you get to live again; if it doesn't work, you're still dead and you don't care that you lost the premiums on the life insurance policy. If you don't sign up, whether or not cryonics works, in the future, you're still dead. In a century or so we'll know the outcome of the experiment, so why not put yourself in the experimental group instead of the control group?[43]

It's a good argument, but cryonics is not a matter of nothing to lose and everything to gain. A life insurance policy purchased in your sixties could set you back between $3,000 and $5,000 per year. If you live another twenty years, that means you've spent between $60,000 and $100,000 on premiums that might have been invested elsewhere, such as in real estate, the stock market, or your family. And might not your money be better spent in extending your life *now* rather than the promised future (unless, of course, you have unlimited disposable income)? If Alcor or one of the other cryonics organizations offered me a free freeze

I'd take it, but in the meantime if I had to bank on one of these technologies I would opt for the extropians and transhumanists because at least they suggest a more protopian approach of change in incremental steps that I can employ starting tomorrow (diet, exercise, lifestyle), and in fact I do that already, as do most people who care about their health and longevity. So let's continue along that path and see how far we can get. Maybe medical breakthroughs in the coming decades will enable my generation to make it into our nineties or low one hundreds in relative physical health and cognitive coherence, and perhaps genetic engineering will allow more and more of us to lead healthy and happy lives into our 120s. But for the immortality meliorists fantasizing about living for centuries or millennia (or forever), remember that the Second Law of Thermodynamics is paramount in the universe, so entropy will get us in the long run, if not the short. As the renowned physicist and astronomer Sir Arthur Stanley Eddington explained in his classic work *The Nature of the Physical World*:

> The law that entropy always increases—the second law of thermodynamics—holds, I think, the supreme position among the laws of Nature. If someone points out to you that your pet theory of the universe is in disagreement with Maxwell's equations—then so much the worse for Maxwell's equations. If it is found to be contradicted by observation—well, these experimentalists do bungle things sometimes. But if your theory is found to be against the second law of thermodynamics I can give you no hope; there is nothing for it but to collapse in deepest humiliation.[44]

I am also skeptical of extrapolating trend lines very far into the future. Human history is highly nonlinear and unpredictable. All those nifty graphs of accelerating technological change may not continue at those rates, nor apply to all biotechnologies. The downsizing of computers from room size to pocket size is one thing; it is quite another to go from pocket size to cell size. The miniaturization of computer chips must one day run up against the limitations imposed by the laws of physics and impede many of the laws of accelerating returns that Kurzweil envisions getting us to forever. Plus, in my opinion, the problems of aging and artificial intelligence are orders of magnitude harder than anyone anticipated decades ago when these fields began. Machine intelligence of a human nature is probably decades away, maybe a century, and immortality is at least a millennium away, if not unattainable altogether. Even techno-optimists such as Jay Cornell and R. U. Sirius (Ken Goffman), authors of the encyclopedic work *Transcendence: Transhumanism and the Singularity*, admit that "the human mind is so complex and subtle, and so rooted in our meat bodies, that successful mind uploading will be far more difficult than many think, and perhaps impossible."[45]

Finally, cryonics, transhumanism, the singularity, and mind uploading all sound *utopian*, the other thread we will explore in the next section of the book on the quest for perfectibility and why, like the quest for immortality, it fails to deliver on its promises.

PART III

ALL OUR
YESTERDAYS
AND TOMORROWS

Hope springs eternal in the human breast:
Man never Is, but always To be blest:
The soul, uneasy, and confin'd from home,
Rests and expatiates in a life to come.
Lo! The poor Indian, whose untutor'd mind
Sees God in clouds, or hears him in the wind:
His soul, proud Science never taught to stray
Far as the solar walk, or milky way;
Yet simple Nature to his hope has giv'n,
Behind the cloud-topp'd hill, an humbler heav'n.

—Alexander Pope, *Essay on Man,* 1734

CHAPTER 9

ALL OUR YESTERDAYS

Progress, Decline, and the Pull of Pessimism

Tomorrow, and tomorrow, and tomorrow,
Creeps in this petty pace from day to day,
To the last syllable of recorded time;
And all our yesterdays have lighted fools
The way to dusty death. Out, out, brief candle!

—Shakespeare, *Macbeth*, act 5, scene 5[1]

"If you had to choose a moment in time to be born, any time in human history, and you didn't know ahead of time what nationality you were or what gender or what your economic status might be," what period would you choose? Would it be the Paleolithic, when people lived in tiny bands and tribes? How about the Neolithic beginnings of farming and the birth of civilization? Maybe you'd opt for ancient Egypt, Greece, or Rome and the creation of our modern political, economic, and military institutions? Perhaps medieval times, with their religious rituals, chivalrous knights, and courtly manners would appeal to you? How about Elizabethan England with Shakespearean plays and the Protestant Reformation, or colonial America and the rights revolution that birthed liberty, equality, and justice? The Industrial Revolution in England and America might prove interesting, as would the surge of inventiveness toward the end of the nineteenth century that gave us the telephone, the automobile, and other life-changing inventions. Or maybe you'd like the peaceful years just before the Great War, or the

Roaring Twenties just after? The 1930s that saw the gathering storm that broke out into the Second World War in the 1940s could prove invigorating, or would you prefer the 1950s, when flag, faith, and family values still mattered?

"You'd choose today," answered the man who posed the opening question in an April 2016 speech delivered in Hannover, Germany—President Barack Obama. "We're fortunate to be living in the most peaceful, most prosperous, most progressive era in human history," he opined, adding that "it's been decades since the last war between major powers. More people live in democracies. We're wealthier and healthier and better educated, with a global economy that has lifted up more than a billion people from extreme poverty."[2] Later that year, in his final speech before the United Nations, Obama piled on the optimism, noting that genetic engineering is leading to cures for diseases that have plagued humanity for centuries, that a young girl in a remote village with a smartphone has access to the entirety of human knowledge, and that those born today will be healthier, live longer, and have more opportunities than anyone in history.[3]

Surely this is political hyperbole, a president spin-doctoring his own term in office into something extraordinary. No, what Barack Obama reflected in his orations is what a number of us have been saying for years—*these* are the good old days.[4] There is no period in history when it would have been better to be alive than today. People who fantasize about a romantic past imagine themselves living in Pharaoh's court, Caesar's palace, Plato's athenaeum, a medieval knight's manor, a king's castle, a queen's château, an emperor's citadel, a cardinal's cathedral. But the cold, hard reality is that 99.99 percent of all the people who ever lived existed in what we would today consider squalid poverty. Even history's "one percent" enjoyed few to none of the luxuries that even an average middle-class Westerner today takes for granted: medical and dental care and public health measures and medicines that enable most people to live into their seventies and eighties; homes with heating and air conditioning, refrigeration, gas or electric stoves, dishwashers, washer/dryer units, swimming pools, gardens, and assorted other creature comforts; more than 10 billion products from which to choose at supermarkets, warehouse outlets, and online stores with cheap home delivery (now even by drones); smart cars with safety and navigation systems and (soon) fully autonomous driving; national and international jet travel allowing almost anyone to go almost anywhere in the world in a matter of hours; wireless communications with anyone anywhere anytime; Internet access to all the world's knowledge for free, with millions of exabytes of information produced every year, the equivalent of about 500 trillion smartphones' worth of digital data. Staggering.

Much of this progress is driven by economics. According to the World Bank, the per capita GDP for the entire world reached $10,000 in 2015, almost double the $5,448 in 2000, 17 times higher than a half century before that, and 100 times higher than the way people lived for the first 99,000 years of our existence as a species, when economists estimate that the annual average income per person was the equivalent of a mere $100 (all in 2015 U.S. dollars).[5] And the rate of prosperity's progress is accelerating. Extrapolating from historical trends, the UC Berkeley economist J. Bradford DeLong has computed that the nineteenth-century Industrial Revolution created a 200 percent increase in per capita income over the previous century, but that the twentieth century witnessed an 800 percent increase over the nineteenth. At the current rate of acceleration, the twenty-first century could witness a 1,600 percent increase over the twentieth. If this happens, it will mean that this century will generate more wealth and prosperity for humanity than in all previous centuries combined.[6] It is a mind-boggling conclusion—almost too good to be true.[7]

What about the flip side of the economic ledger—the poor? Even Jesus gave up hope for them and offered only early entrance to heaven. Until now, that has been about the only thing anyone could do to help the poor, but the twenty-first century will at long last see the end of poverty. To put this into perspective, the economic historian Gregory Clark has computed that "the average person in the world of 1800 was no better off than the average person of 100,000 BC." Today, "modern economies are now ten to twenty times wealthier than the 1800 average." And it isn't just that the rich are getting richer. The poor are as well: "There have been benefits aplenty for the typically wealthy owners of land or capital, and for the educated. But industrialized economies saved their best gifts for the poorest."[8] The economist Max Roser has compiled data to show that in 1820, between 84 percent and 94 percent of the world's people lived in poverty or extreme poverty (defined by the United Nations as earning less than $2.50 a day and $1.25 a day respectively, in 2015 dollars). By 1981 the percentage had dropped to

52 percent, and by 2010 it was at slightly less than 20 percent, or one out of five.[9] But the rate of poverty decline is also accelerating, and at the current rate it will reach zero by around 2035.[10] Imagine that: the end of poverty. Who would have thought it ever possible?

WHY THE PESSIMISM?

Given the reality that there has never been another time in history as good as today, why the doom and gloom heaped upon us by politicians and pundits on both sides of the political aisle? In 2015, for example, a YouGov poll conducted for Radio 4 in England found that 71 percent of respondents said they thought the world was getting worse, with only 5 percent believing that it was getting better.[11] In an unscientific survey I conducted at the end of 2016 I asked my 112,000 Twitter followers their opinion on the state of the world and found that 42 percent said it is getting worse, 31 percent said it is getting better, and 27 percent said it is about the same. And these are the people whom I bombard on a near daily basis with statistics and trend lines like those above. Why does it seem that things are bad and getting worse when in fact they are good and getting better? There are six proximate factors and one ultimate element operating to distort reality in the direction of pessimism:

1. *Relative Inequality.* Economically speaking, even though the poor are getting rich, the rich are getting richer faster, so objective progress feels like relative regress. The pie is increasing in size, so everyone gets a bigger slice, but when the rich's already bigger slice increases in size, the relative amount of wealth accumulates more on the upper end, making the incomes of those in the middle and bottom feel smaller. If I make $100,000 a year and Elon Musk makes $100,000,000 a year and we both see a doubling of our incomes, while I should be thrilled at my newfound fortune of $200,000, if I compare it to Musk's massive $200,000,000, that comparative difference may feel worse, even though I'm better off and no one is worse off for the Tesla and SpaceX CEO's fortune. A survey conducted in 2013 found that people tended to *overestimate* the number of American households earning less than $35,000 a year, believing it to be almost half, when the real number is closer to a third. On the other end, people *underestimated* the number of American households earning $75,000 a year or more, believing it to be less

than a quarter when the real figure is a third. They also overestimated income inequality by double, figuring that the richest 20 percent make about 31 times more than the poorest 20 percent, when the real number is 15.5. The average annual income of the richest 20 percent of Americans is in fact $169,000, whereas survey participants pegged the number at $2,000,000, a nearly twelvefold distortion.[12]

2. *Zero-Sum Thinking.* Our evolved intuitions about economics—our folk economics, or what I call *evonomics*[13]—leads us to see most exchanges as *zero-sum*, or win-lose, in which the gain of one means the loss of another. In the small bands and tribes in which our evolutionary ancestors lived for hundreds of thousands of years, most resources were shared, wealth accumulation was almost unheard of, and excessive greed and avarice were punished. There was no economic growth, no capital markets, no "invisible hand" of the marketplace, and no excessive disparity between rich and poor because everyone was, compared to today, dirt poor. There was no accumulation of wealth because there was no wealth to accumulate. In the *nonzero* world in which we live, the gain of one often means the gain of others, and economic freedom coupled to democratic governance and driven by science, technology, and industry has given us an abundance of food and resources. But our cognition operates as if we are still living in that *zero-sum* land of our evolved economic intuitions, making us distrustful of anyone who has more than we do.

3. *Media Bias toward Bad News and Clickbait Punditry.* News media outlets are far more likely to report bad news than good, simply because that is what they've been tasked to do. Another day in Turkey without a coup goes unreported, but just try to take over that country without the world's media covering it, along with millions of smartphone-carrying citizens video recording every incident. One less child starving to death in Africa goes unnoticed, whereas a drought-induced famine that ravages thousands of children on that continent is documented by NGOs and aid organizations. In the clickbait business of punditry and commentary, bad news generates more hits than good news. An opinion editorial, blog, podcast, or YouTube video titled "The Top 10 Ways Life Has Improved This Century" will never be read, viewed, or downloaded as much as one titled "The Top 10 Worst Things That Happened This Century."

4. *Loss Aversion.* On average, *losses hurt twice as much as gains feel good.* The tennis champion Jimmy Connors put it this way in a 1975 *Sports Illustrated* article suggesting that he might throw a match to a

friend: "C'mon, dammit. Whoever says I tank will get a punch in his chops. I hate losing more than I love winning."[14] In the film *Moneyball*, Brad Pitt's character, Billy Beane, general manager of the Oakland A's, explained the psychology of loss aversion in baseball: "You get on base, we win. You don't, we lose. And I *hate* losing. I hate losing more than I even wanna win."[15] The champion cyclist Lance Armstrong echoed the sentiment in explaining to filmmaker Alex Gibney that he was more motivated not to let cancer—and subsequently other cyclists—defeat him, than he was by the pull of the positive payoffs of winning: "I like to win, but more than anything, I can't stand this idea of losing. Because to me, losing means death."[16]

Behavioral economists have demonstrated experimentally that in order to get someone to take a gamble or to risk an investment, the potential payoff must be about twice the potential loss. To get a person to toss a coin to win or lose $10 (students), or $10,000 (wealthy executives), the payoff has to be greater than or equal to $20, or $20,000. Loss aversion may be found both in gambling behavior and stock market investing. Gamblers, for example, are highly sensitive to losses and tend to follow them with bigger bets, and yet turn conservative when winning by placing smaller bets.[17] The behavioral economist Richard Thaler asked subjects who they would rather be in the following scenario, Mr. A or Mr. B:

> Mr. A is waiting in line at a movie theater. When he gets to the ticket window he is told that as the one hundred thousandth customer of the theater, he has just won $100.
>
> Mr. B is waiting in line at a different theater. The man in front of him wins $1,000 for being the one millionth customer of the theater. Mr. B wins $150.

Remarkably, most people said they would prefer to be Mr. A—they were willing to forgo $50 so as not to feel the pain of not winning $1,000.[18]

5. *Endowment Effect.* We are more willing to invest in defending what is already ours than we are to take what is someone else's, because the motivation to avoid losing what we already have is greater than the motivation of gaining what we don't yet have. For example, dogs will invest more energy and emotion in defending a bone from a challenger than they will in trying to nab some other dog's bone. In people, Richard Thaler tested this effect when he gave subjects a coffee mug said to

be valued at $6 and then asked them what they would sell it for. The average price was $5.25. He then asked another group of subjects how much they would pay for the same mug. The average price was $2.75.[19]

So evolution has wired us to care more about what we already have than what we might possess, and the effect has been found in primates like small-brained capuchin monkeys. In one experiment the monkeys were given tokens they could use as a form of currency for food (grapes and apple slices) that they exchanged with the experimenters, who then manipulated the conditions in such a way that the monkeys had a choice of a 50 percent chance of a bonus of more food or a 50 percent chance of a loss of food they had already earned, with the result that the monkeys were twice as averse to the loss as they were motivated by the gain.[20] Since monkeys, apes, and humans are closely related primates, this finding suggests that loss aversion and the endowment effect evolved in a common ancestor many millions of years ago.

6. *Negativity Bias, or Bad Is Psychologically Stronger than Good.* In a now classic paper titled "Bad Is Stronger Than Good,"[21] the psychologists Roy Baumeister, Ellen Bratslavsky, Catrin Finkenauer, and Kathleen Vohs discovered that across many domains, on average and in the long run, the psychological effects of negative outcomes trump the effects of those that are positive:

- Bad smells elicit far more animated facial expressions than good or neutral odors.[22]

- Bad impressions and negative stereotypes form faster and are more resistant to change than positive ones.[23]

- Memory recall is better for bad behaviors, events, and information than it is for good.[24]

- Negative stimuli have a stronger influence on neural activity than positive stimuli.[25]

- Losing money and friends has a greater impact on people than gaining money and friends.[26]

- Criticism and negative feedback hurt more than praise and positive feedback feel good.[27]

- A study of the emotional content of diaries found that bad events negatively impacted both good and bad moods, whereas good events were limited to affecting only good moods.[28]

- Bad everyday events have a greater impact than good; for example, having a good day does not necessarily lead to a good mood the next day, whereas a bad day often does carry over its consequences into the next day.[29]

- Bad information is processed more thoroughly than is good information; e.g., euphoria after winning the lottery fades quickly compared to the longer-lasting adverse effects of being paralyzed in an automobile accident.[30]

- Traumatic events leave traces in mood and memory longer than good events; e.g., a single childhood traumatic event such as sexual molestation can erase years of positive experiences.[31]

- It is a well-known effect that physical proximity of strangers in a living environment (e.g., college dorms) is a strong predictor of who will become friends, but an even stronger predictor of who will become enemies.[32]

- In an analysis of more than 17,000 psychological research papers, 69 percent of them dealt with negative issues compared to only 31 percent positive,[33] most likely because bad things have a greater impact on human thought and behavior than do good, so there is a more urgent need to understand problems in order to find solutions, and thus the results are more likely to get funded and published.[34]

- Morally bad actions by far weigh more on the moral evaluation of a person by others than morally good actions. In an aptly titled article "Two Rights Don't Make Up for a Wrong," the authors found that "the overall goodness of a person is determined mostly by his worst bad deed."[35] Decades of devoted work for public causes can be obliterated in an instant with an extramarital affair, financial scandal, or criminal act.

In a long review paper covering hundreds of studies, the psychologists not only consistently found that in all domains of life, bad is stronger than good, they were unable to find a *single* counterinstance in which good outdid bad. And it wasn't for lack of trying. "We had hoped to identify several contrary patterns, which would have permitted us to develop an elaborate, complex, and nuanced theory about when bad is stronger versus when good is stronger," Baumeister and colleagues conclude. "It is found in both cognition and motivation; in both inner, intrapsychic processes and in interpersonal ones; in connection with decisions about the future and to a limited extent with memories of the past; and in animal learning, complex human information processing, and emotional responses."[36]

The psychologists Paul Rozin and Edward Royzman identify a similar effect they call a *negativity bias*, in which "negative events are more salient, potent, dominant in combinations, and generally efficacious than positive events."[37] To the numerous examples above of how pessimism trumps optimism, Rozin and Royzman add these:

- Negative events lead us to seek causes more readily than do positive events. Wars, for example, generate endless analyses in books and articles, whereas peace literature is paltry by comparison. Economic recessions and bear markets spawn hand-wringing searches for deep causes, whereas slow and steady economic growth and long-lasting bull markets leave nothing comparable in the explainosphere.

- Picking out an angry face in a crowd is easier and faster to do than picking out a happy face.[38]

- Pain feels worse than no pain feels good. As the philosopher Arthur Schopenhauer put it, "we feel pain, but not painlessness."[39] For the most part, when it comes to the body, "no news is good news." We become aware of our body parts only when something goes wrong. And pain may be felt in more places than pleasure. There are erogenous zones, Rozin and Royzman point out, but no corresponding torturogenous zones.

- We have more words to describe the qualities of physical pain (deep, intense, dull, sharp, aching, burning, cutting, pinching, piercing, tearing, twitching, shooting, stabbing, thrusting, throbbing, penetrating, lingering, radiating, etc.) than we have to describe physical pleasure (intense, delicious, exquisite, breathtaking, sumptuous, sweet, etc.).[40]

- There are more cognitive categories for and descriptive terms of negative emotions than positive.[41] As Leo Tolstoy famously observed in 1875, since elevated to the *Anna Karenina Principle*: "Happy families are all alike; every unhappy family is unhappy in its own way."[42]

- There are more ways to fail than there are to succeed. As Aristotle wrote in the *Nicomachean Ethics*: "For men are good in but one way, but bad in many." It is difficult to reach perfection, and the paths to it are few, but there are many ways to fail to achieve perfection, and the paths away from it are many.

- Empathy is more readily triggered by negative stimuli than positive: people identify and sympathize with others who are suffering or in pain more than they do others who are in a state happier or better off than they.[43] As Jean-Jacques Rousseau remarked in his 1762 book *Émile*: "It is not in the human heart to put ourselves in the place of those who are happier than ourselves, but only in the place of those who are most to be pitied."

- Evil contaminates good more than good purifies evil. As the old Russian proverb says, "A spoonful of tar can spoil a barrel of honey, but a spoonful of honey does nothing for a barrel of tar." In India, members of the higher castes may be considered contaminated by eating food prepared by members of the lower castes, but those in the lower castes do not receive an equivalent rise upward in purity status by eating food prepared by their upper-caste counterparts.[44]

- The gradient for evil is almost always steeper than that of good. For example, in animal learning experiments, an approach-avoidance conflict was presented to rats, but when they reached the goal region they were not only rewarded with food but also punished with a mild shock. This leads to approach-avoidance ambivalence about reaching the end of the runway, so the rats end up vacillating, first toward, then away from the goal. When harnessed to a device to measure their pull strength either toward or away from the goal, psychologists have been able to take a quantitative reading on precisely how strong or weak their ambivalence was. Revealingly, as the rats got closer to the goal box, the strength of both approach and avoidance tendencies increased, but the avoidance gradient was stronger than that of the approach.[45]

- The notorious "one drop of blood" rule of racial classification has its origin in the Code Noir, or Negro Code, of 1685, meant to guarantee the purity of the white race by screening out the tainted blood, whereas, note Rozin and Royzman, "there exists no historical evidence for the positive equivalent of a 'one-drop' ordinance—that is, a statute whereby one's membership in a racially privileged class would be assured by one's being in possession of 'one drop' of the racially superior blood."[46]

To these six proximate factors we can add one deeper ultimate element to get to the heart of why we focus on the bad even when things are demonstratively good, and why we seem to prefer pessimistic and declinist views of history over optimistic ones. That deeper element is evolution.

THE EVOLUTIONARY LOGIC OF PESSIMISM

There's a good reason for the cognitive asymmetry between good and bad: progress is mostly made incrementally and in small steps, whereas regress can easily come about in one colossal calamity. In a complex machine or body, for example, all the parts must consistently work to keep the thing going, but if one part or system fails it can be catastrophic to all the other parts and systems if the machine or organism stops or dies. Stability of the overall system must be maintained, which requires the brain running the system to devote the most attention to threats that could terminate the organism. You live only as long as everything works, so the good news of, say, experiencing yet another day of your heart steadily beating goes unnoticed; but a nonfatal cardiac arrest focuses the mind on this single bad event. And for good reason. There are many ways that things can go south quickly, and this creates an asymmetry between life and death.

Steven Pinker expands on this evolutionary explanation when he notes that in our evolutionary past there was an asymmetry of payoffs in which the fitness cost of overreacting to a threat was less than the fitness cost of underreacting, so we err on the side of overreaction, i.e., pessimism. Pinker places the blame for our evolved pessimism squarely on the shoulders of the Second Law of Thermodynamics, or entropy. In our world—particularly the world in which our ancestors evolved their cognition and emotions that we inherited—entropy dictates that there are more ways for things to go badly than well, so our modern psychology is attuned to a world that was more dangerous in our past than it is today. "The Second Law defines the ultimate purpose of life, mind, and human striving: to deploy energy and information to fight back the tide of entropy and carve out refuges of beneficial order," Pinker notes.[47] The ne plus ultra explanation for entropy can be found on the bumper sticker: SHIT HAPPENS. As such, so-called misfortunes such as accidents, plagues, famine, and disease have no purposeful agency behind them—no gods, demons, or witches intending us evil—just entropy taking its course. Poverty doesn't need an explanation, because it is what you get if you do nothing; wealth needs an explanation, and economic science gives us one.[48]

In a more dangerous world, it paid to be risk averse and highly sensitive to threats, and if things were good, then taking a gamble to improve them a little bit more was not worth the risk of things turning for the worse. To model this phenomenon more precisely, consider an explanation for why we tend to find meaningful patterns in both meaningful and meaningless noise, or what I call *patternicity*. My thought experiment goes like this: Imagine you lived three million years ago on the plains of Africa as a tiny small-brained bipedal primate highly vulnerable to many predators. You hear a rustle in the grass. Is it just the wind or is it a dangerous predator? If you assume that the rustle in the grass is a dangerous predator but it turns out that it is just the wind, you have made a *Type I error*, or a *false positive*—believing something is real when it isn't. You connected the rustle in the grass to the dangerous predator, but in this case there was no connection. No harm. You move away from the rustling sound and become more alert and cautious. But if you assume that the rustle in the grass is just the wind and it turns out to be a dangerous predator, you have made a *Type II error* in cognition, or a *false negative*—believing something is not real when it is. You failed to connect the rustle in the grass to the dangerous predator, and in this case there was a connection, possibly making you the predator's next meal. To avoid making such errors in cognition, why can't you just wait in the grass to collect more data about the rustle in the grass? Because predators don't wait around for potential prey to collect more data about them—that's why they are stealthy and stalk their prey. So the default position is to assume that most rustles in the grass are dangerous predators and not just the wind. In my book *The Believing Brain*, I constructed this formula to model *patternicity*:

$$P = C_{TI} < C_{TII}$$

Patternicity (P) will occur whenever the cost (C) of making a Type I error (TI) is less than the cost (C) of making a Type II error (TII).[49]

This was modeled on the well-established Hamilton's Rule, named for the renowned British evolutionary biologist William D. Hamilton, which states:

$$P = br > c$$

A positive (P) social interaction between two individuals may occur when the benefit (b) of the genetic relatedness (r) exceeds the cost (c) of the social action.

A sibling, for example, may make an altruistic sacrifice for another sibling when the cost of doing so is surpassed by the genetic benefits derived from getting its genes into the next generation through the surviving sibling.

In this context, we might think of *pessimism* as the default position for living in a dangerous world. If it turns out there is no danger, no harm is done and little energy is expended in being pessimistic. If it turns out that there was danger, being pessimistically cautious pays off. In other words, *Assume the worst!* Jared Diamond calls this "constructive paranoia," and in his book *The World Until Yesterday* shows how his approach to risk assessment is practiced by the native New Guineans he had studied for decades.[50] An evolutionary formula might look something like this:

$$P = C_{AW} < C_{AB}$$

Pessimism (P) results whenever the cost (C) of Assuming the Worst (AW) is less than the cost (C) of Assuming the Best (AB).

In this configuration, *pessimism* is a type of *pattern*, a belief about the world in which it paid off for our ancestors to be more negative than positive, and an evolutionary explanation for a worldview in which pessimism trumps optimism. Our minds evolved in that world, not the far safer modern world, so our pessimism can seem misplaced when confronted with the deluge of data showing that optimism—or at least gratitude—would be a more appropriate response. It also explains our longing for the Good Old Days, the idea of decline in our own days, and the yearning for a Golden Age to come, either in heavens above or a heaven on earth.

THE PULL OF THE PAST

Today's pessimism is often accompanied by a longing to return to a mythical time and paradisiacal place. This is not a yearning only of us moderns. The ancient Greeks, for example, believed that they lived in the Age of Iron and that before them were the superior Ages of Bronze, Silver, and Gold, in that Olympian order moving backward in time. This framework was first laid down by the Greek poet Hesiod, a deteriorationist who believed that in that prior Golden Age,

> *Men lived like gods without sorrow of heart, remote and free from toil and grief: miserable age rested not on them; but with legs and arms never failing they made merry with feasting beyond the reach of all evils. When they died, it was as though they were overcome with sleep, and they had all good things; for the fruitful earth unforced bare them fruit abundantly and without stint. They dwelt in ease and peace.*

The Greek Golden Age came to an end when, famously, Prometheus offered fire to humanity, for which he was punished by Zeus, who chained him to a rock where an eagle pecked at his liver for eternity. The decline was furthered when Pandora opened her forbidden box, thereby unleashing evil upon the world.

The Romans picked up the declinist theme when the poet Virgil echoed Hesiod in his characterization of a prior idealized state called Arcadia, in which

> *Fields knew no taming hand of husbandmen*
> *To mark the plain or mete with boundary-line.*
> *Even this was impious; for the common stock*
> *They gathered, and the earth of her own will*
> *All things more freely, no man bidding, bore.*

Therefore, the Greek historian Polybius insisted, *carpe diem* and *"tomorrow, do thy worst, for I have lived today."*

The declinist theme was carried on in Roman times, as in Ovid's *Metamorphoses,* in which the poet declared

The Golden Age was first; when Man, yet new,
No rule but uncorrupted Reason knew:
And, with a native bent, did good pursue.
Unforc'd by punishment, un-aw'd by fear.

Augustus Caesar broke the cycle of doom with his victory over Marc Antony and Cleopatra at Actium, thereby restoring the Golden Age of Rome:

Ours is the crowning era foretold in prophecy;
Born of time, a great new cycle of centuries
Begins. Justice returns to Earth, the Golden Age
Returns, and its first born comes down from heaven . . .

The most famous and persistent paradisiacal fantasy is the mythical Garden of Eden, where humans lived in primordial love and concord with God and nature, as described in Genesis 2:8–9:

The LORD God planted a garden eastward in Eden, and there He put
the man whom He had formed. And out of the ground the LORD God
made every tree grow that is pleasant to the sight and good for food.
The tree of life was also in the midst of the garden, and the tree of the
knowledge of good and evil.

Unfortunately, that audacious autodidact Eve dared to educate herself by partaking of the fruit of that tree of the knowledge of good and evil, thereby bringing about the fall and with it Yahweh's punishment of condemning all humanity to toil in thorn- and thistle-infested fields, to suffer plagues, diseases, accidents, and disasters, and for all future women to undergo unbearable pain in childbirth. By the end of the Bible, St. John the Divine's book of Revelation portrays the terminus of this cycle of history with the return of the messiah and the prelapsarian world of peace and harmony.

THE GOOD OLD DAYS: THEY WERE DREADFUL

Why do we hark back to such mythical Golden Ages? One factor is that we tend to confuse changes in ourselves with changes in our time and culture, an observation backed by data presented in a paper titled "The Ideology of the Good Old Days," in which the psychologists Richard Eibach and Lisa Libby note that as we age, (1) we take on more responsibilities, so we have a greater cognitive burden, (2) we become more vigilant about threats (especially as parents) and more sensitive to errors in youth ("kids these days!"), (3) while at the same time we lose the capacity to process information as quickly as we did when we were younger, and (4) we tend to attribute these changes in ourselves to changes in the external world. "When people fail to realize that these personal changes have heightened their perceptions of threat," Eibach and Libby explain, "they may mistakenly conclude that threats are becoming more prevalent in society."[51] This results in familiar tropes: "Things aren't the way they used to be." "They don't make things like they used to." "It's not like it was in the good old days."

Moral decline in particular has been identified in surveys as what most taxes people's anxieties, many citing the mythical 1950s as their standard of comparison, all contradicted by the data. Rates of volunteering and inflation-adjusted charitable contributions both increased since that decade, the authors note in response to a 2002 poll in which people reported the opposite;[52] crime rates were in the middle of a long plunge in 1999, when 73.7 percent of respondents said they thought crime had increased;[53] just as helicopter parenting became a thing in the 1990s, a majority of survey takers in 2004 said that children were being neglected;[54] a 2003 poll reported that 68 percent of adults thought that teen pregnancy was on the rise, when in fact it had declined by 31 percent since 1991.[55] "People can honestly claim to see more crime, disorder, and immorality in the world today then they did growing up," Eibach and Libby conclude. "However, they fail to realize that they are seeing more of these things because they themselves are different now: now they are worried parents while then they were carefree teenagers, or now they have adult responsibilities while then they were free to explore life's opportunities."

The good old days were when we were young, and every generation feels that way, as Robert Bork noted in his fittingly titled book *Slouching Towards Gomorrah*: "Regret for the golden days of the past is probably universal and as old as the human race. No doubt the elders of prehistoric tribes thought the younger generation's cave paintings were not up to the standard they had set."[56] The longing for a more glorious past and its concomitant grumbling about the present is especially poignant in political rhetoric.[57] In his 1651 political treatise *Leviathan*, Thomas Hobbes noted: "Competition of praise inclineth to a reverence of antiquity. For men contend with the living, not with the dead."[58] For many modern American political commentators, the 1950s provides an ink-blot test for what they think of as the Good Old Days, which the investigative journalist Tina Dupuy notes is especially blinkered inasmuch as they "completely missed the Red Scare, McCarthyism and an equally terrifying polio epidemic. Let's not forget segregation, thalidomide and lobotomies. Also from a conservative perspective, the top tax rate was 70 percent, way higher taxes than we have now."[59]

Indeed, a 2015 poll conducted by the Public Religion Research Institute found that 53 percent of Americans agreed with the latter part of this question: "Since the 1950s, do you think American culture and way of life has mostly changed for the better, or has it mostly changed for the worse?" As Dupuy sardonically notes, "Meaning: More than half of Americans won't simply admit that the Civil Rights Act moved the country forward and *Leave It to Beaver* was absolute dreck. But this idea is a motivator. It makes people believe if we just get back to those principles, like police brutalizing and jailing homosexuals, we can be good once again." In this sense, "The Garden of Eden myth is something to hold on to. To the distressed, the Good Old Days is the perfect *oh shit* handle" (the handle above automobile passenger windows that can be grabbed when the driver takes a tight turn or comes to an abrupt halt). If the 1950s were not the Good Old Days, when were they? It depends on whom you ask, Dupuy notes:

> Of course in the 1950s, groups like the Klan were trying to get the country back to the 1850s. In the 1850s, groups like the Know Nothings, who opposed Catholic immigrants, were trying to get the country back to the 1810s. In the 1600s, the Puritans were trying to get their country back to the Old Testament. We should all go back to a time before nostalgia was used to manipulate people.

When was that? Never. All our yesterdays cast a shadow over all our tomorrows.

CHAPTER 10

ALL OUR TOMORROWS

Utopias and Dystopias in Fiction and in Fact

"A new heaven and a new earth: for the first heaven and the first earth were passed away," we read in Revelation. Cross out "heaven," just keep the "new earth," and you have the secret and the recipe of all utopian systems.

—E. M. Cioran, *History and Utopia*, 1960[1]

In his panoramic illustrated history of imagined places titled *The Book of Legendary Lands*, the late Italian philosopher and novelist Umberto Eco reflected on what lay behind the human desire to live somewhere and some time else: "It seems that every culture—because the world of everyday reality is cruel and hard to live in—dreams of a happy land to which men once belonged, and may one day return."[2] In his famous "I Have a Dream" speech, the Reverend Dr. Martin Luther King Jr. imagined such a land where his children would grow up to live in a country "where they will not be judged by the color of their skin but by the content of their character" and where ultimately the "rough places will be made plain, and the crooked places will be made straight."[3]

As Umberto Eco reminds us, however, there's a darker side to these dreams, as what resulted from the quest for Ultima Thule, said to be a distant perfect land located beyond the boundaries of the known world, "a land of fire and ice where the sun never set," and "the cradle of civilization was supposedly the north, and from there the mother races had spread out toward the south," in Eco's description. This Ultima Thule was said to be the origin of the Aryan race, from which all others degenerated. In his 1885 book *Paradise Found*, Boston University rector William F. Warren located this earthly paradise at the North Pole, contending that the first inhabitants were beautiful and long-lived but then deteriorated as they moved south. Many occultists also speculated on where this first and purist race originated. In her 1888 book *The Secret Doctrine*, for example, Madame Blavatsky ventured that the perfect race came from a polar continent that extended from Greenland to Kamchatka. The Order of the New Temple was founded in 1907 by Jörg Lanz, an Austrian racial theorist who preached that "inferior races" should be sterilized and deported to Madagascar so as not to contaminate the pure Aryan race. In 1918 the Thule Gesellschaft (Thule Society) was

Figure 10-1. The Emblem of the Thule Gesellschaft

The Sanskrit symbol for "good fortune," a hooked cross called the *svastika*, was adopted by the Thule Society, which was devoted to the historical and anthropological search for the origin of the superior Germanic race.

established, and for its emblem adopted the Sanskrit symbol for "good fortune," a hooked cross called the *svastika* (figure 10-1). In 1935 a former chicken farmer instituted the Society for Research and Teaching of the Ancestral Inheritance, devoted to the historical and anthropological search for the origin of the superior Germanic race. His name was Heinrich Himmler, and he went on to became the Reichsführer of the Nazi *Schutzstaffel* (SS) and the titular head of the Reich's *die Endlösung der Judenfrage*—the final solution to the Jewish problem. Such is the power of myth when put into action.

The heavens on earth we are considering in this book are not restricted to spiritual spaces or physical firmaments; they also include numerous attempts over the millennia to create a literal heaven on earth in the form of *utopias*—places where imperfect humans can strive for perfectibility: personal, political, economic, and social. The dark mirror to utopias are *dystopias*, real and imagined failed social experiments, repressive political regimes, and overbearing economic systems.

UTOPIAS AND DYSTOPIAS IN FICTION

In the imagination of fiction, utopian worlds have been envisaged in a staggering array of forms of what life would be like if only . . . the impossibility of which is found in the definition of *utopia*—"no place"—Sir Thomas More's neologism coined for his 1516 work that launched the modern genre.[4] His island of Utopia is located somewhere in the Atlantic and reflects the era's New World discoveries unfolding in the decades prior (see figure 10-2). Typical of the genre in its more communal form, on Utopia there is no private property, there are no locks on doors, goods are stored in warehouses from which people take only what they need, and everyone works an essential job in agriculture, metalworking, masonry, carpentry, weaving, and other basics of life. People dress alike and do not seem to have many wants beyond rudimentary necessities; there is no unemployment, medical care is free, privacy is not desired, and everyone works a six-hour day, made possible by the fact that every household has two slaves. There is much scholarly debate over why More wrote *Utopia* and to what extent he was commenting on the deficiencies of his own society,[5] but it seems clear enough that he was opining on the impossibility of achieving societal perfection, as

Figure 10-2. The Island of Utopia

Woodcut map by Ambrosius Holbein for the 1518 third edition of Thomas More's *On the Best State of a Republic and of the New Island Utopia,* or simply *Utopia.* From the Hulton Archive, British Library collection. Reproduced courtesy of Getty Images.

evidenced by the book's additional neologisms: Polyleritae ("Muchnonsense"), Macarenses ("Happiland"), the river Anydrus ("Nowater"), and the character Hythlodaeus ("Dispenser of Nonsense").

Social commentary is in fact what most of these works are really about. Consider the lost continent of Atlantis, a mythic utopia that has been projected to have been in the Mediterranean, the Atlantic (the Canaries or the Azores being remnants), Iceland, Sweden, the Caribbean, or the Pacific (between South America and Antarctica, or somewhere between Australia, New Guinea, and the Solomon and Fiji islands). The evidence for the lost continent was apparently washed away when it vanished beneath the waves, but that hasn't quelled the imagination of utopia hunters. In his 1954 book *Lost Continents: The Atlantis Theme in History, Science, and Literature*, the science fiction author L. Sprague de Camp counted 216 different "Atlantists," only 37 of whom concluded that Atlantis was imaginary or allegorical, with the rest convinced the real lost continent could be found.[6] In 1989 the French underwater treasure hunter Pierre Jarnac tallied more than five thousand book titles about Atlantis, and this was pre-Internet. In his 2012 book *Atlantis: In the Textual Sea*, Andrea Albini reported that more than 23 million web pages were devoted to the imagined utopian continent.[7]

There is in fact no point in searching for Atlantis, because Plato made up the story as a social commentary on Athens and a warning to his fellow Athenians to pull back from the precipice that war and wealth were taking them over. In the *Timaeus*, Plato's dialoguist, Critias, explains that Egyptian priests told the Greek wise man Solon that his ancestors once defeated a mighty empire located just beyond the Pillars of Hercules (usually identified by Atlantologists as the Strait of Gibraltar). "This vast power gathered into one, endeavored to subdue at a blow our country and yours and the whole of the region within the straits; and then, Solon, your country shone forth, in the excellence of her virtue and strength, among all mankind." But afterward, "there were violent earthquakes and floods; and in a single day and night of misfortune all your warlike men in a body sank into the earth, and the island of Atlantis in like manner disappeared in the depths of the sea." Atlantis, in fact, is not a utopian place to be emulated, but a warning about what happens when a society is corrupted by excessive belligerence and avarice. When that happened to the Atlanteans Zeus called forth the other gods to his home, "and when he had gathered them there he said . . ." The text abruptly ends there, but Plato's point was made. Figure 10-3 is one of many portrayals of Atlantis and its destruction, this from the artist Thomas Cole, rendered in 1836.

The fodder for Plato's imagination came from his experiences growing up at the terminus of Athens's Golden Age, brought about, in part, by the costly wars against the Spartans and Carthaginians. He visited cities such as Syracuse, featuring numerous Atlantean-like temples, and Carthage, whose circular harbor was controlled from a central island,

Figure 10-3. The Destruction of Atlantis

The Course of Empire: Destruction, painted in 1836 by Thomas Cole (1801–1848), represents the ruin of Atlantis. Collection of the New-York Historical Society, object #1858.4. Reproduced with permission.

as in Atlantis. Earthquakes were common: when Plato was fifty-five, one leveled the city of Helice, only forty miles from Athens, and, most tellingly, the year before he was born, an earthquake had flattened a military outpost on the small island of Atalantë. Plato wove historical fact into literary myth, as he did in his more famous work, *The Republic*, in this case to warn how a utopia can become corrupted into a dystopia. As he explained: "We may liken the false to the true for the purpose of moral instruction." The utopian myth is the message.

Dystopias have a later genealogy, the word coined in 1868 by the utilitarian philosopher John Stuart Mill in a debate before the House of Commons, when he denounced the government's Irish land policy: "It is, perhaps, too complimentary to call them Utopians, they ought rather to be called dys-topians, or cacotopians. What is commonly called Utopian is something too good to be practicable, but what they appear to favour is too bad to be practicable." The odd reference is to the original antonym to utopia, coined by the founder of utilitarianism, Jeremy Bentham, but its clumsy neologism—cacotopia (wicked place)—didn't take.

If a utopia is an imaginary place where everything is as good as possible, a dystopia is an imaginary place in which everything is as bad as possible.[8] As the historian of utopias Howard Segal notes, "one person's or one society's utopia may be another's anti-utopia or 'dystopia.'"[9] More's Utopia, for example, would be a dystopia for most of us, resembling as it does North Korea. "If someone takes the license to wander far from his own district and is caught without the pass issued by the supreme magistrate," More narrates the law in his idealized society, "if he dares to do so a second time, he is condemned to slavery." Ayn Rand's novel *Atlas Shrugged* is another utopian/dystopian work, which opens with a society in decline as a result of the actions of her hero, John Galt, who leads a strike by the "men of the mind," forcing civilization to collapse into chaos, out of the ashes of which the heroes resurrect an Atlantis on earth. As the book's two heroes fly over the shattered ruins of a once great civilization now darkened into a charred landscape, Rand's heroine, Dagny Taggart, proclaims, "It's the end." No, Galt rejoins, "It's the beginning." Dystopia is followed by utopia. Heavens on earth.

UTOPIAS AND DYSTOPIAS IN FACT

Utopias are the idealized visions of a perfect society; *utopianisms* are those ideas put into practice.[10] Here is where the trouble begins. People act on their beliefs, and if you believe that the only thing preventing you and/or your family, friends, and clan from going to heaven or achieving heaven on earth is someone else or some other group, and that heavenly visage includes eternal life in the hereafter or infinite goodness in the here and now, their evilness knows no bounds for what is permitted to end it. From homicide to genocide, the murder of others in the name of some religious or ideological belief accounts for the high body counts in history's conflicts, from the Crusades, the Inquisition, witch hunts, and religious wars of centuries past to the world wars and genocides of the past century. As the political philosopher John Gray expressed it in his book *Black Mass*: "Utopias are dreams of collective deliverance that in waking life are found to be nightmares."[11]

The failures of the experimental new societies of the nineteenth century, such as New Harmony in Indiana (1825–29), Brook Farm in

Massachusetts (1841–46), Oneida Community in New York (1848–81), and Octagon City in Kansas (1856–57), were relatively harmless because they did not employ force and violence upon dissenters, but instead tried to sell everyone on the benefits to come. "Plodding on their weary march of life, Association rises before them like the mirage of the desert," wrote John Humphrey Noyes, a utopian socialist and the founder of the Oneida Community. "They see in the vague distance magnificent palaces, green fields, golden harvests, sparkling fountains, abundance of rest and romance; in one word HOME—which is also HEAVEN." Noyes's vision of heaven included communal sharing and free love, directed by and benefiting mostly himself (in the usual custom of cults), and failed after his death. But at least no one was killed in the quest (and they had a lot of sex).

There were nearly a hundred such planned utopian communities in the United States in the mid-nineteenth century, some religious, some secular, and all utopian in the sense that they were doomed to fail from the start. Fruitlands, for example, a utopian community founded by disgruntled Mormons in 1843, lasted only seven months when their strict vegetarian diet and money-free economy collided with the reality of their first New England winter. Ralph Waldo Emerson articulated the problem all such utopian communities face when describing Charles Fourier's utopian socialist schemes: "Fourier skipped no fact but one, namely life."[12] A friend of Emerson named George Ripley, a Unitarian minister, founded the Brook Farm Institute of Agriculture and Education in 1841 in Roxbury, Massachusetts. Grounded in the utopian socialism of Fourier and known simply as Brook Farm, this community also practiced free love. The Brook Farm experiment came to an end six years later when the communal living structure was consumed by fire in 1847 and there were no funds or motivation to rebuild it.

Such intentional communities, as they are sometimes called, continue today, albeit in different forms. There is Tamera, for example, founded by a German psychoanalyst named Dieter Duhm (along with a theologian named Sabine Lichtenfels) in 1978 in the Black Forest of Germany, then reestablished in 1995 in Portugal with a goal of becoming "a self-sufficient, sustainable and duplicable communitarian model for nonviolent cooperation and cohabitation between humans, animals, nature, and Creation for a future of peace for all."[13] Well, who could be against such lofty goals? In reality, Tamera operates more as an experimental research center than a self-sustaining community, like a college campus embedded in a stable society. Their Love School will sound familiar: its aim is to realize love without jealousy, sexuality without fear, faithfulness with love and desire for others, and "new pathways" in partnerships; in other words, free love. I encountered a similar experimental community in a 2017 visit to the famous Freetown Christiania in the middle of Copenhagen, Denmark, an "autonomous neighborhood" of about 760 residents (630 adults, 130 children) that was founded in 1971 in abandoned military barracks by trespassing homeless people who encamped there and whom the authorities largely ignored (I lectured there in 2012 but it was night so I couldn't see much). Most live in the graffiti-festooned deserted buildings and makeshift shacks scattered about the property, with lots of gardens, dirt paths, and waterways connecting the fifteen disparate subcommunities (see my photographs in figure 10-4). I was not allowed to photograph "Pusher Street," and for good reason: the traditional cash crop that has long supported the community—cannabis in various forms—is not technically legal. (You can see it in the distance to my left in the selfie.) "The objective of Christiania is to create a self-governing society whereby each and every individual holds themselves responsible over the well-being of the entire community," contends the mission statement. "Our society is to be economically self-sustaining and, as such, our aspiration is to be steadfast in our conviction that psychological and physical destitution can be averted."[14] Self-sustainability through companies like Christiania Bikes and Helena Jewelry may be a goal, but the place seemed indistinguishable from the 1960s California hippie communes of my youth. In fact, I was told by my hosts that the camera-shy drug dealers are in actuality part of an organized drug cartel that fixes prices and processes profits from mostly hashish and marijuana sales. Although this is something of an anomaly, on August 31, 2016, two police officers and a civilian were shot by an armed drug dealer when law enforcement moved in to shut down their operations. Other raids have turned up weapons, ammunition, bulletproof vests, homemade firecrackers, chrysanthemum bombs, and—the currency of all underground economies—cash.

What brings down most such communities is a litany of human and societal foibles well articulated by the investigative journalist Alexa Clay in her examination of why most human organizations fail:

Malarial infested swamps, false prophecy, sexual politics, tyrannical founders, charismatic con-men, lack of access to safe drinking water, poor soil quality, unskilled labour, restless dreamer syndrome, land not suitable for farming: all sensationalise the rocky history of intentional communities. But the more relevant drivers that cause many communities to unravel sound more like the challenges afflicting any organisation today: capital constraints, burn-out, conflict over private property and resource management, poor systems of conflict mediation, factionalism, founder problems, reputation management, skills shortage, and failure to attract new talent or entice subsequent generations.[15]

Most such intentional communities are unsustainable without outside support from patrons, benefactors, and donors. A more sustainable business model for an intentional community is the Esalen Institute in Big Sur, California, a retreat center for human potential that I have visited many times. It sits on a stunning bluff overlooking the Pacific Ocean where paying customers take courses and workshops, get massages and sit in natural hot spring spas, meditate and do yoga, eat healthful organic meals, and hike the local trails. Prices are lower than at the five-star resorts in nearby Monterey, but one can easily rack up thousands of dollars in a multiday visit unless you're willing to sleep in a sleeping bag in a communal area. People who live there are either paid staff or volunteers who earn their keep in the kitchen, maintaining the gardens and grounds, or facilitating workshops and other activities, and an annual fundraising drive helps keep the multimillion-dollar budget in the black. The facilities are spartan but the views spectacular (see figure 10-5). Still, it is better thought of as a retreat than a utopia.[16]

Worse, but not socially catastrophic, are utopian cults, such as the Order of the Solar Temple in Switzerland in the 1980s and 1990s, forty-eight members of whom were murdered by various means and then had their bodies torched in a fire. Then there was the Heaven's Gate UFO cult founded by Marshall Applewhite and Bonnie Nettles in 1975, claiming to have arrived via UFO from another dimension above human. Members sold their possessions, lived in isolation, and traveled in pairs so as not to be unduly influenced by outsiders. They practiced living in dark rooms to simulate space travel, and lived an ascetic lifestyle with minimal material possessions in a monastic-like order. Sex was considered sinful in all relationships, and six male

Figure 10-4

Impressions of Freetown Christiania by the author in a 2017 visit to the "autonomous community" in the middle of Copenhagen, Denmark. (A) One of the abandoned military barracks now occupied by residents of Christiania. (B) One of the many makeshift homes. Bicycles are a common form of transportation on the grounds. (C) Graffiti on one of the occupied buildings indicating the community's main cash crop. (D) Pusher Street to the left of the author in this selfie—photographs are not allowed therein, as selling hashish and marijuana is not legal.

Figure 10-5. The Esalen Institute

Figure 10-5. The Esalen Institute

A human potential intentional community with a sustainable business model enables this facility to continue indefinitely, thereby setting it apart from most utopian communities. Photographs by the author: (A) a basic room, (B) spectacular views, (C) gardens, (D) kitchen.

members even voluntarily underwent castration to curb their temptations. By the mid 1990s the group was earning its keep through web services, registering the business name "Higher Source" and creating the website heavensgate.com (still in operation by surviving members). One night in early 1997, the group heard a radio broadcast of Art Bell's late-night show *Coast to Coast AM*, a regular purveyor of conspiracy theories with little or no facts to support them. This episode featured a guest who regaled the listening audience with the incredible story that the much talked about appearance of Comet Hale-Bopp in the night sky disguised a secret—a UFO was trailing close behind it on a secret mission to Earth. Heaven's Gate members came to believe that this was the vehicle that would take them to what they called The Evolutionary Level Above Human (TELAH), a "physical, corporeal place" where they would live forever in pure bliss. Never mind that there wasn't a shred of evidence of such an extraterrestrial vehicle behind the comet, much less that its mission was to rescue a couple dozen members of a New Age cult living in San Diego. How they would get to the UFO was another matter entirely, and a deadly one at that. On March 25, 1997, Applewhite persuaded 39 of his followers to expedite the journey by imbibing a deadly cocktail of phenobarbital and vodka, and for good measure they voluntarily pulled plastic bags over their heads for self-asphyxiation. Authorities found them the next day in their rented San Diego home lying supine on their beds, dressed in identical black shirts, sweat pants, and Nike shoes and sporting armband patches that read HEAVEN'S GATE AWAY TEAM.

The deadliest of the utopian cults was the Reverend Jim Jones's Jonestown, 918 members of which committed suicide or were murdered on November 18, 1978, while Jones cajoled them to make the ultimate choice between the collapsing community they built in the jungles of Guyana and the utopian life that awaited them once they passed over:

> I'm glad it's over. Hurry, hurry, my children . . . No more pain . . . Death is a million times preferable to ten more days of this life. If you knew what was ahead of you, you'd be glad to be stepping over tonight . . . This is a revolutionary suicide. It's not a self-destructive suicide . . . If we can't live in peace, then let's die in peace. Take our life from us. We laid it down. We got tired. We didn't commit suicide, we committed an act of revolutionary suicide protesting the conditions of an inhumane world.[17]

Many orders of magnitude more catastrophic were the grand twentieth-century experiments in utopian socialist ideologies as manifested in Marxist/Leninist/Stalinist Russia (1917–89), Fascist Italy (1922–43), and Nazi Germany (1933–45), which were large-scale attempts to achieve political, economic, social, or racial perfection, resulting in tens of millions of people murdered by their own states or killed in conflict with other states perceived to be blocking the road to paradise. The Marxist theorist and revolutionary Leon Trotsky expressed the utopian vision in a 1923 pamphlet:

> The human species, the coagulated *Homo sapiens*, will ... become immeasurably stronger, wiser and subtler; his body will become more harmonized, his movements more rhythmic, his voice more musical. The forms of life will become dynamically dramatic. The average human type will rise to the heights of an Aristotle, a Goethe, or a Marx. And above this ridge new peaks will rise.[18]

This unrealizable goal led to such bizarre experiments as those conducted by Ilya Ivanov, whom Stalin tasked with crossbreeding humans and apes in order to create "a new invincible human being." When Ivanov failed to produce the man-ape hybrid, Stalin had him arrested, imprisoned, and exiled to Kazakhstan.[19] As for Trotsky, once he had power as one of the first seven members of the founding Soviet Politburo, he established concentration camps for those who refused to join in this grand utopian experiment, ultimately leading to the gulag archipelago that killed millions of Russian citizens who were also believed to be standing in the way of the imagined utopian paradise to come. When Trotsky's own theory of Trotskyism opposed the theory of Stalinism, the dictator had him assassinated in Mexico in 1940.

Following the Second World War, in the second half of the twentieth century, revolutionary Marxism in Cambodia, North Korea, and numerous states in South America and Africa led to murders, pogroms, genocides, ethnic cleansings, revolutions, civil wars, and state-sponsored conflicts, all in the name of establishing a heaven on earth that required the elimination of recalcitrant dissenters. All told, some 94 million people died at the hands of revolutionary Marxists and utopian communists in Russia, China, North Korea, and other states, a staggering number compared to the 28 million killed by the fascists.[20] When you have to murder people by the tens of millions to achieve your utopian dream, you have succeeded only in producing a dystopian nightmare.

In their book *Why Not Kill Them All? The Logic and Prevention of Mass Political Murder*, Daniel Chirot and Clark McCauley draw the parallels between Christian and Marxist eschatology. In Marx's prelapsarian world there was no private property, no classes or class division, no alienation between employees and employers. But then came the fall and the invention of private property, leading to exploitation and economic classes of haves and have-nots. This sinful world can last only so long before the apocalypse in which "a final, terrible revolution will wipe out capitalism, alienation, exploitation, and inequality." The chosen people in this eschatology are, naturally, the Communists, with Marx as their savior. Then there's Hitler's Thousand Year Reich, another imagined eschatological scenario with Christian parallels, as Chirot and McCauley elaborate:

> It was not an accident that Hitler promised a Thousand Year Reich, a millennium of perfection, similar to the thousand-year reign of goodness promised in Revelation before the return of evil, the great battle between good and evil, and the final triumph of God over Satan. The entire imagery of his Nazi Party and regime was deeply mystical, suffused with religious, often Christian, liturgical symbolism, and it appealed to a higher law, to a mission decreed by fate and entrusted to the prophet Hitler.[21]

Humans are not perfectible because no such thing as perfection exists, either individually or collectively, but the belief that it does leads to infallibility of method and no means to correct the inevitable mistakes that arise when designing a perfect society for an imperfect species.[22] Utopias tend to fail as a result of a flawed theory of human nature in which collective ownership, communal work, authoritarian rule, and a command-and-control economy collide with individualism, the desire for autonomy, and natural differences in ability, leading to inequalities of outcomes and imperfect living and working conditions that utopias committed to equality of outcome cannot tolerate.[23] As one of the original citizens of Robert Owen's New Harmony community in Indiana explained it: "We had tried every conceivable form of organization and government. We had a world in miniature. We had enacted the

French revolution over again with despairing hearts instead of corpses as a result. It appeared that it was nature's own inherent law of diversity that had conquered us."[24] In his 1964 Republican presidential nomination speech, Barry Goldwater identified one of the fundamental problems of utopias when they collide with the human desire for power:

> Those who seek absolute power, even though they seek it to do what they regard as good, are simply demanding the right to enforce their own version of heaven on earth. And let me remind you, they are the very ones who always created the most hellish tyrannies. Absolute power does corrupt, and those who seek it must be suspect and must be opposed. Their mistaken course stems from false notions of equality, ladies and gentlemen. Equality, rightly understood, as our founding fathers understood it, leads to liberty and to the emancipation of creative differences. Wrongly understood, as it has been so tragically in our time, it leads first to conformity and then to despotism.[25]

The utopian quest for perfect happiness was exposed as the flawed goal that it is by George Orwell in his 1940 review of *Mein Kampf*, when the journalist and critic observed that Hitler "has grasped the falsity of the hedonistic attitude to life." Whereas most progressives and liberals after the Great War had come to believe that people want only a life free from pain and strife, Orwell noted that Hitler understood that people want more than just "comfort, safety, short working-hours, hygiene, birth-control and, in general, common sense; they also, at least intermittently, want struggle and self-sacrifice." On the broader appeal of fascism and socialism, Orwell added that Fascism, Nazism, and Socialism (as practiced under Stalin) are sounder theories of human nature, inasmuch as people need challenges and goals, not just pleasure. "Whereas Socialism, and even capitalism in a more grudging way, have said to people 'I offer you a good time,' Hitler has said to them, 'I offer you struggle, danger, and death,' and as a result a whole nation flings itself at his feet." Orwell added, as a warning, "we ought not to underrate its emotional appeal."[26]

The ultimate problem of utopian logic begins with a utilitarian calculus in which everyone will live in perfect harmony once we get rid of any dissenters who don't see as clearly as the collective. We know from research on the now famous Trolley Problem that most people would be willing to kill one person in order to save five. Here's the setup: You are standing next to a fork in a railroad line with a switch to divert a trolley car that is about to kill five workers on the track. If you pull the switch it will divert the trolley down a side track where it will kill one worker. If you do nothing the trolley kills the five. What would you do? Most people say that they would pull the switch.[27] If even people in Western (presumably enlightened) countries today agree that it is morally permissible to kill one person to save five, imagine how easy it is to convince people living in autocratic states with utopian aspirations that they should, say, kill a thousand to save five thousand, or eliminate a million so that five million may prosper. What's a few zeros when we're talking about infinite happiness and eternal bliss? This is what Thomas Paine warned against in his 1795 *Dissertation on First Principles of Government*: "He that would make his own liberty secure, must guard even his enemy from oppression; for if he violates this duty, he establishes a precedent that will reach to himself."[28]

THE IDEA OF UTOPIAN PROGRESS

The long history of the idea of progress is documented by J. B. Bury in his 1920 classic *The Idea of Progress*, and more recently by Robert Nisbet in his 1980 *History of the Idea of Progress*, showing how these ideas mostly surface as either a counter to declinist trends or to bolster belief in a utopian state—both religious and secular—progress toward which feels almost millennarian. Bury distinguishes between progress that results from human aims and will to bring about, say, "liberty, toleration, and equality of opportunity" and ideas such as "fate, providence or personal immortality," the latter treating progress as if it were a force of nature separate from human action.[29] Nisbet reiterates that "the idea of progress holds that mankind has advanced in the past—from some aboriginal condition of primitiveness, barbarism, or even nullity—is now advancing, and will continue to advance through the foreseeable future." Nisbet also notes, however, that the idea of progress is what lay beneath the twentieth-century totalitarianisms, left and right, so the idea itself may or may not result in desirable outcomes. As Cardinal Newman observed: "Men will die for a dogma who will not even stir for a conclusion." Nisbet adds: "The idea of the slow, gradual, inexorable progress of mankind to higher status in knowledge, culture, and moral estate is a dogma."[30]

That may be, but in practice those communities and societies who embrace the idea of progress that succeed are of the *protopian* type: incremental progress in steps toward *improvement*, not perfection. As the futurist Kevin Kelly describes his neologism: "Protopia is a state that is better today than yesterday, although it might be only a little better. Protopia is much harder to visualize. Because a protopia contains as many new problems as new benefits, this complex interaction of working and broken is very hard to predict."[31] In *The Moral Arc* I showed how protopian progress best describes the monumental moral achievements of the past several centuries: the attenuation of war, the abolition of slavery, the reduction of torture and the death penalty, and the advancement of universal suffrage, liberal democracy, civil rights and liberties, same-sex marriage, and animal rights.[32] These are all examples of protopian progress in the sense that they happened one small step at a time. And there is nothing teleological or inevitable about it, because progress is not itself a force of nature; it is a result of human action.

THE IDEA OF DYSTOPIAN DECLINE

People act on their beliefs, and if you believe that civilization is in decline and that you can do something to stop it, that can be a recipe for violence. Case in point: the "Olympic Park Bomber" Eric Rudolph's explanation for what motivated him to set off explosive devices in the middle of the 1996 Olympic games in Atlanta. "The decision to act was the result of many years of my being confronted with the decline of Western civilization and the realization that only radical action would slow or halt this decline."[33] The following year, Rudolph bombed an abortion clinic and a lesbian bar in Atlanta, and in 1998 he bombed an abortion clinic in Birmingham, Alabama. After his capture in June 2003 it was learned that he was a member of the Christian Identity movement, an extremist organization that believes Jews are Satanic and blacks are subhuman.[34] He told law enforcement officials that he wanted to combat abortion and the "homosexual agenda" he believed to be the product of a "rotten feast of materialism and self-indulgence," and to protect "the integrity of American society" and "the very existence of our culture." As he elaborated: "To place the homosexual relation-ship along side of the model and pronounce it to be just as legitimate a lifestyle choice is a direct assault upon the long term health and integ-rity of civilization and a vital threat to the very foundation of society."[35]

The motivation to respond violently to the perceived decline of civi-lization also drove the Harvard-educated mathematical prodigy turned domestic terrorist, the Unabomber Ted Kaczynski, who between 1978 and 1995 targeted individuals for death with bombs delivered through the mail, killing three and leaving twenty-three others maimed. In his rambling fifty-page manifesto, *Industrial Society and Its Future*,[36] pub-lished in the *New York Times* and elsewhere in hopes that some reader would recognize the degenerationist rhetoric of the author (Kaczynski's brother did, which is how he was caught), Kaczynski opined that the Industrial Revolution was "a disaster for the human race." Although technologies increased the life expectancy of people living in these industrialized countries, they have "destabilized society, have made life unfulfilling, have subjected human beings to indignities, have led to widespread psychological·suffering (in the Third World to physical suffering as well) and have inflicted severe damage on the natural world. The continued development of technology will worsen the situa-tion." Modern Americans, he griped, are "decadent leisured aristo-crats" who are "bored, hedonistic, and demoralized," nothing more than "domesticated animals." Speaking in the first-person plural, or occasionally as "FC" (Freedom Club), like many revolutionaries before him, Kaczynski advocated revolution, not necessarily a violent sudden one, but "a relatively gradual process spanning a few decades." He went on to outline "the measures that those who hate the industrial system should take in order to prepare the way for a revolution against that form of society." This was not intended to be a political revolution, Kaczynski continued, not an overthrow of governments, but an over-turning of "the economic and technological basis of the present soci-ety." This revolution is to be followed by the introduction of a new ideology "that opposes technology and the industrial society . . . so that when and if the system collapses, the remnants will be smashed beyond repair, so that the system cannot be reconstituted."

In a 2010 postscript to the manifesto written in prison, Kaczynski reflected on the many commentaries he had read about his writings, some of which accused him of unoriginality and compared him to radical environmentalists: "Many radical environmentalists and 'green' anarchists talk of revolution, but as far as I am aware none of them have shown any understanding of how real revolutions come about, nor do they seem to grasp the fact that the exclusive target of revolution must be technology itself, not racism, sexism, or homophobia."[37] The cover of Kaczynski's book (figure 10-6) captures the bleak perspective such dystopian pessimism evokes. Although his declinism

Figure 10-6. The cover of the Unabomber Ted Kaczynski's manifesto

was driven, in part, by mental illness, his views are shared by a great many people.

Intellectuals are just as susceptible to the pull of a paradisiacal past followed by a declinist history, albeit without the violence. For at least a century and a half, public intellectuals and academic scholars have been predicting the imminent collapse of Western civilization, even as the ideals and institutions that ensure its success have grown: science and technology, reason and Enlightenment humanism, democracy and universal franchise, property rights and the rule of law, free enterprise and free trade, and the rights of individuals expanded to include all humans and even members of other sentient species. You would think academics and intellectuals—the very people who promote such values—would be singing their own praises for such progress, but no, they're gloomier than ever. An Amazon book scan of the word string "the coming crisis" produces titles like *The Coming Financial Crisis* (2015), *The Global Water Crisis* (2008), *The Coming Bond Market Collapse* (2013), *Get Prepared Now! Why a Great Crisis Is Coming and How You Can Survive It* (2015), *ISIS, Iran and Israel: What You Need to Know about the Current Mideast Crisis and the Coming Mideast War* (2016), *The Coming Oil Crisis* (2012), *Coming Climate Crisis?* (2012), *The Coming Economic Armageddon* (2010), *The Coming Inflation Crisis* (2014), *The Coming Famine* (2011), *Rising Sea Level and the Coming Coastal Crisis* (2012), and fifty-five more pages of titles in a similar vein.

"The decline and fall of" string returns, naturally, Gibbon's *The Roman Empire*, but also *Byzantium, the Habsburg Empire, the Ottoman Empire, the British Empire, the British Aristocracy, the Japanese Empire, the Soviet Empire, the Roman Church, Radical Catholicism, the American Republic, Democracy in 21st Century America, American Growth, Truth in Bush's America,* and *the West,* to cite just a few from the first ten of a hundred pages of such doom. "The rise and fall of" generates 11,857 titles, starting with William L. Shire's classic *Third Reich* and continuing with the rise and fall of *Ancient Egypt, Classical Greece, Alexandria, Carthage, the Roman Empire, the House of Caesar, the House of Medici, the British Empire, Communism, the Cherokee Nation, American Growth, American Business, the Constitution, Society, Nations, Empires,* and *the Great Powers.* The only two positive titles I found were on the rise and fall of *Violent Crime* and *Slavery.*

In the late 1980s I devoured Paul Kennedy's 1987 *The Rise and Fall of the Great Powers* and assigned it for a class I taught on the history of war. I completely bought into his thesis that America had reached "imperial overstretch" and was soon to go the way of the British Empire and its Pax Britannica. "The task facing American statesmen over the next decades," Kennedy warned, would be "to 'manage' affairs so that the relative erosion of the United States' position takes place slowly and smoothly."[38] This was just three years before the collapse of the Soviet Union, which was not even on Kennedy's radar, in keeping with nearly everyone else in the West, including Soviet experts. Not that democracy was spared the doomsayers' prognostications. *The Twilight of Democracy*, *The Democracy Trap*, *Democracy on Trial*, *Giving Up on Democracy*, *The Frozen Republic*, *The Selling of America*, *The Bankrupting of America*, *The Endangered American Dream*, and *Who Will Tell the People* (without even a question mark) were just a few titles from the 1990s, all during the exceptionally prosperous administration of the democratic centrist Bill Clinton.

The first decade of the twenty-first century saw a continuation of the trend, as evidenced in two excerpts from popular works written by leading public intellectuals.[39] The first is the opening passage of three-time U.S. presidential candidate and political analyst Patrick Buchanan's 2008 book *Churchill, Hitler, and the Unnecessary War*, a history and social commentary that made the *New York Times* bestseller list and was the cover story for the June 23 issue of *Newsweek*:

> All about us we can see clearly now that the West is passing away. In a single century, all the great houses of continental Europe fell. All the empires that ruled the world have vanished. Not one European nation, save Muslim Albania, has a birthrate that will enable it to survive through the century. As a share of world population, peoples of European ancestry have been shrinking for three generations. The character of every Western nation is being irremediably altered as each undergoes an unresisted invasion from the Third World. We are slowly disappearing from the Earth.

The second comes from the epilogue of the Harvard historian Niall Ferguson's 2006 epochal history, *The War of the World*, also presented in a widely viewed PBS documentary series:

A hundred years ago, the West ruled the world. After a century of recurrent internecine conflict between the European empires, that is no longer the case. A hundred years ago, the frontier between West and East was located somewhere in the neighbourhood of Bosnia-Herzegovina. Now it seems to run through every European city. That is not to say that conflict is inevitable along these new fault lines. But it is to say that, if the history of the twentieth century is any guide, then the fragile edifice of civilization can very quickly collapse even where different ethnic groups seem quite well integrated, sharing the same language, if not the same faith or the same genes.

This historical revisionism aims to explain the decline of the West by reconfiguring World War II as "the unnecessary war" instead of "the good war" (Buchanan), by combining the two world wars into one long ethnic and economic conflict that could have been avoided had England left Germany alone (Ferguson), and to demonstrate the moral equivalency between the Axis and the Allies in the outbreak and conductance of a war whose waging probably failed to help those who most needed it (Ferguson). Although the United States emerged from the conflict as the world's foremost military, economic, and financial power, the war was seen by these authors as a long-term setback for European culture and Western civilization.

THE BLOOD AND SOIL ROMANCE

Such declinism has a long history, chronicled by the historian Arthur Herman in his magisterial *The Idea of Decline in Western History*,[40] much of which was deeply entwined with what declinists believed accounted for the *rise* of the West in the first place—family, race, ethnicity, and nation.[41] Call it *the blood and soil romance* (from the German slogan *Blut und Boden*), an anti-Enlightenment, Romantic pastoral fantasy of racial, national, and cultural purity grounded in ethnicity and geography, a longing for a return to racially pure peoples, more rigid top-down controls over the unwashed and ignorant masses, a reversion to a constitutional monarchy, perhaps, or a benevolent dictatorship, where flag, faith, and family hold centrality and where everyone knew their place in the rigid class system and those at the top called the shots.

The groundwork for the blood and soil romance was laid a century before, starting with the racial pessimism of Count Joseph Arthur de Gobineau, the nineteenth-century French aristocrat who claimed to trace his genetic heritage to Normandy's Viking invaders, the same lineage that gave the world William the Conqueror. His mother said she was descended from an illegitimate son of King Louis XV, while his father fought on the royalist side in the French Revolution. In 1835 the young Gobineau left Normandy for Paris, where he encountered disdain among intellectuals for the philistines—the bourgeois industrialists and merchants who saw science, technology, and industry as the future drivers of progress, not to mention the ticket to upward mobility. Whereas Enlightenment scientists looked back to the Middle Ages as a time of superstition to be jettisoned, the intellectual circles Gobineau traveled in looked at it with romantic nostalgia, a time before blood and soil were contaminated with modernity. The political revolutions of 1848 divided intellectuals who were disturbed by the violence, such as Alexis de Tocqueville, from those who thought they were not violent enough, such as Karl Marx and Friedrich Engels, who famously declared at the end of their *Manifesto of the Communist Party* that the revolutionary Communists "openly declare that their ends can be attained only by the forcible overthrow of all existing social conditions. Let the ruling classes tremble at a Communistic revolution. The proletarians have nothing to lose but their chains. They have a world to win. Working Men of All Countries, Unite!"[42]

For Gobineau, however, the class-based revolutions were contrary to his own self-identified aristocratic *noblesse de race*, which he saw as the driving force behind civilization, leading him to pen his most famous work in 1853, *The Essay on the Inequality of the Human Races*. "The great events—the bloody wars, the revolutions, and the breaking up of laws—which have been rife for so many years in the States of Europe, are apt to turn men's minds to the study of political problems," he began, griping that while "the vulgar consider merely immediate results, and heap all their praise and blame on the little electric spark that marks the contact with their own interests, the more serious thinker will seek to discover the hidden causes of these terrible upheavals." After "passing from one induction to another," Gobineau arrived at "the conviction that the racial question overshadows all other problems of history, that it holds the key to them all, and that the inequality of the races from whose fusion a people is formed is enough to explain the whole course of its destiny."[43]

The birth of racial science can be traced back to the early part of the nineteenth century and the birth of anthropology, when Georges Cuvier and Johann Friedrich Blumenbach classified humans into Oriental/ Mongol (yellow), Negroid/Ethiopian (black), and Caucasian (white). Debate ensued among anthropologists as to whether the different races constituted varieties of one species (*monogenists*) or different species (*polygenists*). Monogenists generally believed that all present human races were the product of a slow deterioration from the perfect creation in the Garden of Eden. This degeneration, however, was not believed to have been equal among races. In the racial ranking of most nineteenth-century intellectuals, some races fell further from the brood creation than others—blacks most and whites least, with "red" Egyptians and American Indians (and others of varying shades of nonwhite) in between. Polygenists, on the other hand, did not believe in a single womb from which all races were born. Rather, they suspected that there were multiple progenitor "Adams" from which each race descended, currently represented by these separate racial types. "On the basis of skeletal and cranial evidence, polygenists insisted that blacks were physically distinct and mentally inferior," notes the historian George Stocking in his panoramic history of racial anthropology;

> On the basis of the racial representations on "ancient Egyptian monuments" they argued that races had remained unchanged throughout the major portion of human history; on the basis of the mortality of whites in tropical areas they hypothesized that different races were aboriginal products of different "centers of creation" and could never fully "acclimate" elsewhere; on the basis of anecdotal evidence they asserted that the hybrid offspring of blacks and Europeans were only partially interfertile.[44]

As the monogenist-polygenist debate raged on, the topic turned to cultural differences between the races, leading naturally to the proposal that some races were more advanced than others with, as customary for the era, the purer white race leading to progress while racial mixing led to decline. The linguistic study of language similarities led some scholars to suggest that Europeans may once have shared a common language, along with a common culture, an idea that was co-opted by racial scientists to propose that there was once an original pure race—Aryan—that, like the Atlantis and Golden Age myths before, came to represent the nineteenth-century version of the superior Lost Civilization from which modern civilizations had declined. "I may thus lay it down, as a universal axiom, that the hierarchy of languages is in strict correspondence with the hierarchy of races," Gobineau concluded, following the usual racial ranking: "The negroid variety is the lowest, and stands at the foot of the ladder. The animal character, that appears in the shape of the pelvis, is stamped on the negro from birth, and foreshadows his destiny. His intellect will always move within a very narrow circle." Moving up the ladder, Gobineau notes that the "yellow man has little physical energy, and is inclined to apathy . . . his desires are feeble, his will-power rather obstinate than violent . . . he tends to mediocrity in everything; he understands easily enough anything not too deep or sublime . . . The yellow races are thus clearly superior to the black. Every founder of a civilization would wish the backbone of his society, his middle class, to consist of such men. But no civilized society could be created by them." Civilized society, of course, was the product of "white peoples . . . gifted with an energetic intelligence . . . a feeling for utility, but in a sense far wider and higher, more courageous and ideal, than the yellow races . . . a greater physical power, an extraordinary instinct for order . . . a remarkable, and even extreme, love of liberty . . ." and on and on in this vein for pages. "A society is great and brilliant only so far as it preserves the blood of the noble group that created it," he concluded. But now, racial degeneration had led to a decline of Western civilization as "the blood of the civilizing race is gradually drained away." The mid-century revolutions Gobineau witnessed were emblematic of the struggle between what was left of the prototype Aryan-German aristocracy, which he saw himself descended from, and the new bourgeois classes clambering for change. "The white species will disappear henceforth from the face of the earth," although humans "will not quite have disappeared but will have completely degenerated . . . deprived of strength, beauty, and intelligence."[45] Figure 10-7 captures the sentiment of the day in this 1895 painting by Hermann Knackfüss, inspired by Gobineau and the racial science of the day, presented by Emperor William II to the Russian tsar, depicting "the Yellow Peril" (Buddha on the right) encroaching on Europe, titled *Nations of Europe, Join in the Defense of Your Faith and Your Homes!*

Figure 10-7. The Threat of Racial Mixing

An 1895 painting by Hermann Knackfüss portrays the Archangel Michael warning the peoples of Europe to beware of threats by "the Yellow Peril" (note the Buddha on the right). Reprinted in *Harper's Weekly*, one of the most popular publications of the time.

One fan of Gobineau was the German composer and polemicist Richard Wagner, who told his wife that Gobineau was "my only true contemporary." Wagner introduced Gobineau's racialist theories on the decline and fall of civilizations to his son-in-law Houston Stewart Chamberlain, an Englishman living in Germany whose 1899 book, *The Foundations of the Nineteenth Century*, set the stage for the pan-Germanic movement of the early twentieth century, most particularly the völkisch anti-Semitism of Nazi racial philosophy. Chamberlain bemoaned the decline of English society, telling Wagner that it was primarily the fault of the Jews. Germany's loss in the First World War cemented his belief that his adopted country, like that of his ancestral England, was now under "the supremacy of the Jews." He also railed against the Weimar Republic, which he called a *Judenrepublik*, a Jewish republic. Chamberlain's declinism, however, was tempered when, in January 1921, a colleague wrote to him about "an Austrian worker, a man of extraordinary oratorical talents and an astonishingly rich politi-cal knowledge who knows marvelously how to thrill the masses."[46] After he met him, Chamberlain came to call this man "Germany's savior."

Adolf Hitler, in fact, read Chamberlain's biography of Wagner, and he drew heavily from the racial theorist for his own ideas about racial purity, one of which was that for the Germanic peoples to survive, the Jews would have to be removed from German society. Chamberlain wrote to Hitler after their meeting in 1923: "My faith in Germandom has not wavered for a moment, though my hopes were—I confess—at a low ebb. With one stroke you have transformed the state of my soul. That Germany, in the hour of her greatest need, brings forth a Hitler—that is proof of her vitality."[47] The vitality of race purity had taken on a new threat from the Jews, which Chamberlain believed were a hybrid race of Bedouins, Hittites, Syrians, and Aryans during the period of the Old Testament. Such race pollution, as Chamberlain and the neo-Gobinians saw it (Gobineau himself was not an anti-Semite), led to capitalism, liberal humanitarianism, and "Jewish science," all of which led to a struggle between Aryan Teutons and the Jews. (Jesus, in this bizarre story, was Aryan, not Jewish.)

Another admirer of Richard Wagner's music and philosophy was a precocious philologist named Friedrich Nietzsche, who, after attending a live performance of the overtures to *Die Meistersinger* and *Tristan und Isolde*, proclaimed, "My every fiber, every nerve vibrates to this music."[48] The sentiments were mutual despite the generational age difference between the two men, a gap closed by their shared belief in the decay of modern civilization and their desire to stop it through Romantic heroism led by "men of redemption . . . selected individuals who are equipped for great and lasting works," such as writers, artists, philoso-phers, and composers like Richard Wagner. In his 1872 *The Birth of Tragedy*, Nietzsche argued that the Apollonian spirit of reason and self-control had overcome the Dionysian spirit of art and creativity in the history of civilization, leaving us with "Alexandrian man" who is "equipped with the greatest forces of knowledge" but whose creative spirit had been squelched. Europeans, he grumbled, had become Alexandrian men, but in Wagner's operas Nietzsche found redemption for the decadent civilization around him.

Nietzsche later changed his mind about Wagner's music, concluding that it, too, was decadent, but his later works continued the theme of the decline of European civilization as a result of the loss of the "will to power." "Wherever the will to power declines in any form there is also a physiological regression, a decadence," he orated in *The Anti-Christ*.[49] In *The Will to Power* he declared that history is a dialectical struggle between "those poor in life, the weak" and "those rich in life, the strong."[50] Civilization itself, Nietzsche concluded in *Beyond Good and Evil*, is the product of "men of prey who were still in possession of unbroken strength of will and lust for power, [who] hurled themselves on weaker, more civilized, more peaceful races." Who were these men of prey? It was Gobineau's Aryans, of course, as Nietzsche channeled the decline theorist in his narrative account of the "blond beast," and Christianity, science, and liberal humanitarianism were enslaving the Aryan aristocracy. "For some time now," he wrote in *On the Genealogy of Morals*, "our whole European culture has been moving as toward a catastrophe."[51]

As madness enveloped his mind, Nietzsche penned his only work of fiction, a Wagnerian tribute to "the last man," in *Thus Spake Zarathustra*, in which the Persian religious prophet Zoroaster (Zarathustra in the narrative) symbolizes the vitalism of the ancient Aryan religion that divides the universe into light and dark, life and death. It is here that Nietzsche expands on the theme of the *Übermensch*, subsequently put to music by Richard Strauss with its unforgettable opening trumpet sequence (aptly used by Stanley Kubrick in *2001: A Space Odyssey*), and famously pronounces, "God is dead," calling himself the Antichrist. "Since the old God is abolished, I am prepared to rule the world."[52]

Two authors were profoundly influenced by all these cultural threads woven into the tapestry of the blood and soil romance. The first was Oswald Spengler, educated in philosophy and natural science at the University of Berlin in the first decade of the new century, after which he had a revelation that he was witness to a "world-historical shift" in national identity and cultural destiny, a struggle between the liberal West with its Enlightenment values and Germany with its Romantic ideals. The original title of his magnum opus was *Conservative and Liberal*, but in a Munich bookshop window he glimpsed the title *The Decline of Antiquity*, which gave him inspiration for his new title, *The Decline of the West*. The work was published in 1918, on the eve of Germany's defeat in the First World War. Treating history like a force of nature separate from the individuals peopling the past, Spengler borrowed from his Romantic influences the metaphysical notion of *Volksgeist*, "spirit of the people" (or "national character"), which he saw as a determining force of historical development. "Each culture has its own new possibilities of self-expression, which arise, ripen, decay and never return," he wrote, analogizing civilizations with organisms: "Every culture has its childhood, youth, manhood, and old age."[53] Civilizational old age came to Europe in the nineteenth century, as "amorphous and dispirited masses of men, scrap-material from a great history," meandered pointlessly through the great cities of Europe. What Western culture needed to reverse the decline were "adventurers, self-styled Caesars, seceding generals and barbarian kings." "In our Germanic world the spirits of Alaric and Theodoric will come again" to crush "the dictatorship of money and its political weapon, democracy. The sword will triumph over money."

The other author echoed these recurrent themes of "struggle," "decay," and "decline," and even met Spengler at the Bayreuth Festival, where Richard Wagner and Friedrich Nietzsche had convened half a century before. "They stood at the confluence of two deep and swift channels running through the German cultural landscape," writes Arthur Herman, "one flowing from Gobineau to Wagner and Houston Chamberlain, the other from Nietzsche and his radical nationalist followers."[54] All of these threads were woven into a tapestry of monumental consequence in a book that too few people read and fewer still took seriously. A few excerpts:

> Everything that we admire on this earth today—science and art, technology and inventions—is only the creative product of a few peoples and originally perhaps of one race.
>
> All great cultures of the past perished only because the originally creative race died off through blood poisoning.
>
> Blood mixture and the resultant drop in the racial level is the sole cause of the dying out of old cultures; for men do not perish as a result of lost wars, but by the loss of that force of resistance which is contained only in pure blood.
>
> The Aryan is the Prometheus of mankind from whose bright forehead the divine spark of genius has sprung at all times, forever kindling anew that fire of knowledge which illumined the night of silent mysteries and thus caused man to climb the path to mastery over the other beings of this earth. Exclude him—and perhaps after a few thousand years darkness will again descend on the earth, human culture will pass, and the world turn to a desert.

Those who want to live, let them fight, and those who do not want to fight in this world of eternal struggle do not deserve to live.

The mightiest counterpart to the Aryan is represented by the Jew.

The title of the book is *Mein Kampf,* its author Adolf Hitler.[55]

Thus it is that what the blood and soil romance envisioned as a utopian society (figure 10-8) collapsed into a dystopian nightmare.

Figure 10-8. The Blood and Soil Romance

The *Blut und Boden* (Blood and Soil) emblem on the 4th Reichsbauerntag in Goslar in 1936. Photograph courtesy of Süddeutsche Zeitung Photo, DIZ Dokumentations und Informationszentrum Munchen GmgH, and the Image Works.

All such utopias are premised on a vision of a past that never was and a projected future that can never be, a heaven on earth turned to hell.

UTOPIA, RACE, AND THE RISE OF THE ALT-RIGHT NATIONALISM

The blood and soil romance is not just a historical curiosity of a time long past. Although Western civilization has made remarkable moral progress over the past century, utopian fantasies about political nationalism, economic protectionism, anti-immigration, and especially race and racial purity were lying dormant for decades and have recently arisen in the form of the alt-right, emboldened in 2016 by the appointment of Stephen Bannon to serve as chief strategist to the administration of President Donald Trump. Before joining the Trump team, Bannon was executive chair of the far-right news website Breitbart, which in 2016 Bannon described as "the platform for the alt-right."[56]

The alt-right movement is still fragmented and on the fringe, but it is working to unify its disparate followers through the white nationalist think tank National Policy Institute (NPI) and the website AltRight .com, both run by Richard Spencer, who has been called a white supremacist, a label he rejects. Nevertheless, Spencer advocates for a homeland for a "dispossessed white race," promotes "a new society, an ethno-state that would be a gathering point for all Europeans," decries the "deconstruction" of European culture, and has called for "peaceful ethnic cleansing."[57] According to its website, the NPI "is an independent organization dedicated to the heritage, identity, and future of people of European descent in the United States, and around the world." Rejecting the individualism of classical liberalism, identity in a group is what counts. "Who are you?" asks Spencer in a slick promotional video for the organization. "I'm not talking about your name or occupation. I'm talking about something bigger. Something deeper. I'm talking about your connection to a culture, a history, a destiny, an identity that stretches back and flows forward for centuries."

Unfortunately, Spencer bemoans, "today we seem to have no idea who we are. We're rootless. We've become wanderers." Also rejecting the inclusiveness of Western culture and seemingly ignoring the historical reality that everyone in America not of native descent is an immigrant, Spencer complains that America has "become a country for everyone and thus a country for no one. It's a country in which we ourselves have become strangers." Who are we? "We aren't just white. We are part of the peoples' history, spirit, and civilization of Europe. This legacy stands before us as a gift, and as a challenge."[58] Where have we heard it all before? The "spirit" of a people, a nation, a race is the foundation of the anti-Enlightenment blood and soil racial romance.

Just before the 2016 presidential election, Spencer elaborated on his beliefs and what the alt-right stands for at a press conference (at which the movement's logo was unveiled—see figure 10-9): "I don't think the best way of understanding the alt-right is strictly in terms of policy. I think metapolitics is more important than politics. I think big ideas are more important than policies." He added that "race is real, race matters, and race is the foundation of identity. You can't understand who you are without race."[59] As the name implies, the alt-right is not just anti-left but an alternative to mainstream conservatism, which has shifted socially leftward for decades in response to the movements championing civil rights, women's rights, gay rights, and animal rights. Most conservatives today are more liberal than most liberals were in the 1950s. To those conservatives who have made this shift, Spencer and his alt-right supporters apply the descriptor *cuckservative*—a portmanteau of "cuckold" (the archetypal clueless schmuck) and "conservative." According to George Michael, a criminal justice professor who has studied right-wing movements for decades, a cuckservative is "a conservative sellout who is first and foremost concerned about abstract principles such as the U.S. Constitution, free market economics, and individual liberty. Instead, the alt-right is more concerned about concepts such as nation, race, culture, and civilization."[60] A cuckservative is a conservative who has not only sold out to liberal values, but worse, is weak and ineffectual, an accommodationist, a bipartisan compromiser, and a RINO (Republican In Name Only).[61] Cuckservatives on the right may be as evil as their libtard (another portmanteau of "liberal retard") counterparts on the left.

According to an article posted on AltRight.com titled "What is the Alt Right?" the movement "challenges both the mainstream Left and Right, which it views as essentially two versions of the same Liberal ideology." They reject classical liberals and libertarians who favor lim-

Figure 10-9. The Alt-Right Logo

The alt-right logo was introduced at a press conference in Washington, D.C., in September 2016 by Richard Spencer, president of the National Policy Institute.

ited government, cast off warmongering conservatives with ambitious foreign policy plans, and discard Republicans focused on lowering taxes and growing the economy. Alt-righters are most closely aligned with paleoconservatives, "in their cautious approach to foreign policy and advocacy of traditional values." They praise the aforementioned Patrick Buchanan and laud his pessimistic book *The Death of the West*, and they identify their major intellectual influences as the previously discussed declinist philosophers Friedrich Nietzsche, Martin Heidegger, Thomas Mann, and Oswald Spengler. "Of particular interest are Spenglerian theory of civilizational decline, Nietzschean emphasis on aesthetics and temporal cycles of eternal return, and Schmittian concept of the Political."[62] The pessimistic pull of the past continues today.

The politics of "identity" is central to the alt-right and is in keeping with the utopian blood and soil romance that identity is wrapped up in race—the pure white race, naturally. Even though they acknowledge that historically there have been "cultural communities" in America, such as the Italians and Irish, along with "regional identities" such as the "cowboy culture" in the western states, "the majority of Americans of European descent simply self-identify as 'white.'" And "whereas African, Asian, Hispanic, and other minorities see themselves as fairly coherent communities, with their own demands and institutions, Americans of European descent do not have such organizations and representatives."[63] Until now. Enter the alt-right.

In his analysis, George Michael is careful to separate Donald Trump from the alt-right movement, which was around long before the billionaire's candidacy was announced and would still be on the rise even if he had lost the election. Nor is it restricted to America. "Not unlike the Brexit referendum over the summer of 2016, Trump's startling victory confirms that there is a rising tide of nationalism in the West," Michael concludes. "The increasing popularity of Marine Le Pen could soon lead to a nationalist government in France, which like England, might opt out of the European Union." Yes, at the alt-right press conference, Richard Spencer celebrated Trump's election victory by exclaiming "Hail Trump! Hail our people! Hail victory!" But Michael notes, "To date, however, Trump has eschewed explicit race-mongering" and "has promoted a form of civic nationalism that emphasizes 'America first,'" not white first. In fact, Michael points out that Mitt Romney garnered the same percentage of white votes as Trump, and although Trump's "rhetoric was often construed as impolitic on the campaign stump, he nevertheless reached out to all Americans irrespective of race, gender, sexual orientation, or creed. In fact, he was the first major Republican presidential candidate in many years to have actually made a serious effort to attract African-American voters."[64] Trump even threw his support behind the LGBTQ community, announcing on the popular CBS news program *60 Minutes* that he supports marriage equality and that, on the heels of the terrorist attack on the gay nightclub in Orlando, Florida, he promised to protect LGBTQ Americans along with their rights. And, memorably, PayPal cofounder Peter Thiel—an openly gay man—was invited to speak at the Republican convention.[65] (Although it should be noted that Trump revoked Obama's executive order about workplace discrimination, which would have protected LGBTQ workers, and his stance about transgender people in the military drew the ire of many in the LGBTQ community.) Whatever else one may think about Trump, these are hardly the words and actions of an alt-right racialist demagogue.

In fact, the roots of the alt-right movement go back decades to the Aryan Nations, White Aryan Resistance, National Alliance, and World Church of the Creator, the latter of which preached RAHOWA— Racial Holy War against ZOG, the "Zionist Occupation Government." Backed by a rising militia movement that was moved to action after government overreach led to the disastrous standoffs at Ruby Ridge and Waco, the domestic terrorist bombing in Oklahoma City by Timothy McVeigh in 1995 sent the movement underground. But it didn't die. In June 2008, for example, I attended a conference in Costa Mesa, California, organized by the Institute for Historical Review (which was then a leading Holocaust revisionist organization) and hosted by its director, Mark Weber. The subject was their enduringly favorite passions and pastimes: Hitler and the Nazis, Jews and the Holocaust, and World War II and the decline of the West. In my interview with him at the conference I asked Weber what, precisely, is in decline. "First and foremost, there is a dysgenic trend," he asserted. "The average intelligence level is falling. Everywhere the most educated and cultured peoples are having the least number of children. Music, architecture, and art are in decline. There's a general discordance in culture." By contrast, he said, "A healthy society is cohesive." What does that mean? I pressed him. "Ethnicity and race," Weber responded. By "ethnicity," I continued, do you mean the shared beliefs of a people, as in a common religion? "No," Weber snapped. "Iraqis, for example, share a common religion, but their society is not cohesive. I mean racial or genetic cohesion." For example? I queried. "The Danes are reportedly the happiest people on earth," Weber rejoined. "Certainly a key factor in that regard is the Danes' racial-ethnic cohesion." But, I countered, Americans are a racially diverse society and we are incredibly successful, arguably the wealthiest and most successful nation in history. Weber had an immediate comeback: "The most significant fact of America's history and legacy is that it was settled by Europeans."

In a follow-up telephone interview, I asked Weber about the counterfactual scenario he outlined in his talk at the conference in which he speculated on what might have happened if Britain and France had not declared war against Germany, and the Axis nations had succeeded in obliterating Soviet Communism. What does he think would have happened and what would Europe look like today? It would have resulted in an Axis-dominated Pax Europa, Weber conjectured, a culturally dynamic, socially prosperous, politically stable, economically sound, and technologically advanced culture. "A victorious National Socialist Germany probably would have carried out a space exploration program far more ambitious than that of the United States or the USSR. It would have developed an extensive continent-wide transportation and communications network, an exemplary environmental policy, a comprehensive health care system, and a conscientious eugenics program."[66] Conscientious eugenics? Weber also told me that *Mein Kampf* has recently emerged as a bestseller in certain countries. In Turkey, for example, the people "are turning to Hitler and his philosophy as a viable option to the other failed social experiments." According to Weber, the twentieth century saw the domination of four political systems: communism, theocracy, liberal democracy, and national socialism. Communism is dead. Theocracies are archaic. Liberal democracy—particularly with the decline of America's reputation in the global community—is rapidly falling out of favor. That leaves national socialism.

Although Weber and his Institute for Historical Review have faded from public attention in recent years, the most prominent of the Holocaust revisionists, the British writer David Irving, claims to be experiencing something of a resurgence of interest in his work after he lost his libel lawsuit against Deborah Lipstadt and largely disappeared after the judge's final pronouncement against him, in which he found that "Irving has for his own ideological reasons persistently and deliberately misrepresented and manipulated historical evidence; that for the same reasons he has portrayed Hitler in an unwarrantedly favourable light, principally in relation to his attitude towards and responsibility for the treatment of the Jews; that he is an active Holocaust denier; that he is anti-Semitic and racist and that he associates with right wing extremists who promote neo-Nazism."

One of the defense's expert witnesses, the renowned World War II historian Richard Evans, accused Irving of lying about history and backed up the accusation with enough documentation to fill a book, titled *Lying about Hitler*.[67] Nevertheless, Irving told the *Guardian* in early 2017, "Interest in my work has risen exponentially in the last two or three years." Teenagers in America allegedly find his many lectures on YouTube "and these young people tell me how they've stayed up all night watching them." What are these kids curious about? "They get in touch because they want to find out the truth about Hitler and the Second World War. They ask all sorts of questions. I'm getting up to 300 to 400 emails a day. And I answer them all. I build a relationship with them."[68] I wonder if any of them came across the doggerel Irving sang to his young daughter whenever "half-breed children" were wheeled past them (recorded in Irving's diary, read into the court record,[69] and now all over the internet):

> I am a Baby Aryan
> Not Jewish or Sectarian;
> I have no plans to marry an
> Ape or Rastafarian.

In an unintentionally humorous but revealing moment during the trial, Irving slipped up one day and addressed the judge not as "Your Honor" but as "Mein Führer."[70]

The blood and soil romance of today's alt-right racialism can also be found in the "1488ers," the revolutionary white nationalists who contend (and hope) that racial strife will destroy America and thereby set the stage for the resurgence of white nationalism. The moniker comes from the fourteen-word credo of David Lane: "We must secure the existence of our people and a future for white children." The number 88 stands for the reduplicated eighth letter of the alphabet—HH— signifying "Heil Hitler."[71]

Shortly after the 2016 election, Richard Spencer said that Trump's victory was "the first step, the first stage towards identity politics for white people."[72] It is an illuminating comment inasmuch as it echoes the identity politics practiced by the left for decades, most notably through political correctness and most visually on college campuses. As the foreign affairs professor Walter Russell Mead noted in early 2017:

> The growing resistance among white voters to what they call "political correctness" and a growing willingness to articulate their own sense of group identity can sometimes reflect racism, but they need not always do so. People constantly told that they are racist for thinking in positive

terms about what they see as their identity, however, may decide that racist is what they are, and that they might as well make the best of it. The rise of the so-called alt-right is at least partly rooted in this dynamic.[73]

The obsession on the left with classifying people as members of a race, creed, color, gender, sexual orientation, national origin, ancestry, religion, physical or mental disability, medical condition, marital status, and the like has backfired inasmuch as treating people not as individuals to be judged by the content of their character but by the color of their skin or membership in any of these many collectives is a form of collectivism that ultimately *increases* racism, bigotry, and bias. What may have started out as well-intentioned actions at curbing prejudices and changing thoughts with the goal of making people more tolerant has metamorphosed into thought police attempting to impose totalitarian measures that result in silencing dissent of any kind. In this sense, the alt-right is an understandable reaction to what has come to be called the *regressive left*, which is perhaps best characterized here as the *alt-left*. In the words of the *Wall Street Journal* editor Sohrab Ahmari in his book *The New Philistines*:

Having been told for decades that the promise of universal rights is a lie, that group identity is all there is to public life, that the Western canon is the preserve of Privileged Dead White Men, and that identitarian warfare is permanent, many in the West have taken up their own form of identity politics. There is logic to their demand for validation. When culture only rewards the assertion of group identity (black, female, queer, etc.), the silent majority will want its slice of the identitarian pie. They can do identity politics, too: it's called white nationalism.[74]

What goes around comes around.

PART IV

MORTALITY

AND MEANING

Heaven and hell are within us, and all the gods are within us. This is the great realization of the Upanishads of India in the ninth century B.C. All the gods, all the heavens, all the worlds, are within us. That is what myth is.

—Joseph Campbell, *The Power of Myth*, 1988

CHAPTER 11

WHY WE DIE

The Mortal Individual and the Immortal Species

> To grunt and sweat under a weary life,
> But that the dread of something after death,
> The undiscovered country from whose bourn
> No traveller returns, puzzles the will
> And makes us rather bear those ills we have
> Than fly to others that we know not of?
> Thus conscience does make cowards of us all.
>
> —Shakespeare, *Hamlet*, act 3, scene 1

How long would you like to live? Eighty? A hundred? Two hundred? How about five hundred years? Can you imagine what it would be like to live a thousand years? Me neither.

In surveys that ask such questions most people say that they would not want to carry on much past the current average life expectancy. It's another example of the *status quo bias*, or our emotional preference for whatever we are accustomed to,[1] and our personal life expectations are yoked to those of our generation's life expectancy. According to a 2013 Pew Research Center poll of 2,012 American adults, for example, 60 percent said that they would not want to live past the age of ninety, while another 30 percent said they would prefer to cash out by age eighty. And these findings were consistent regardless of income, belief (or not) in an afterlife, and (in some cases) even anticipated medical advances. When it was proposed that "new medical treatments [could] slow the aging process and allow the average person to live decades longer, to at least 120 years old," a slight majority (51 percent) said that they would *not* personally want such treatments, and that it would be "fundamentally unnatural" and "a bad thing for society."[2]

The "it wouldn't be natural" objection to radical life extension is gainsaid by a simple thought experiment. If you were given a death sentence of, say, tomorrow, would you want to live one more day in order to get your affairs in order and to tell everyone you love how you feel about them? Of course you would. How about one more week? Definitely. Another month? Absolutely. One more year? Well, there are more things I'll bet you'd like to do, so sure you would. Another decade? That would give you time to travel and perhaps even take up a new career, so certainly. At some point I might find the time horizon in which you'd say "that's enough," but fast-forward to the day before that date and we're back to the cycle of wishing for one more day, week, month, year, decade . . . Unless you are terminally ill and in such pain and misery that one more week or month would be manageable only through massive doses of morphine, at no point is a reasonably healthy and happy individual realistically likely to be willing to check out early just because "it wouldn't be natural" to continue. As for society, let the nihilists and the cynics fall on their swords. I'll take another sunrise and sunset, thank you.

The Pew findings bear this out. Respondents were more likely to favor life extension if they are younger, believe that future medical treatments would provide a higher quality of life, if they could still be productive by working longer, if they wouldn't be a strain on our natural resources, if older people were not seen as a problem for society, and if living longer did not result in debilitating diseases and disabilities. What is *natural* is for healthy, happy, and productive people to desire to continue living and loving for as long as they remain healthy, happy, and productive. More and more, people are ignoring the aspiration reflected in the Who's 1960s rock anthem *My Generation*: "I hope I die before I get old." Although the band's mercurial drummer Keith Moon died at the age of thirty-two from the then customary drug overdose, the Who's aging front men Pete Townshend and Roger Daltrey are still touring half a century later.

WHY DO WE AGE AND DIE?

Toward the end of his life, reflecting on the "Topic of Cancer" (that would ultimately kill him) for his *Vanity Fair* column, Christopher Hitchens gave as good an answer as anyone to his self-directed rhetorical query:

> To the dumb question "Why me?" the cosmos barely bothers to return the reply: Why not?[3]

This leads us to a deeper question: Why do we have to die at all? Why couldn't God or Nature endow us with immortality?

The answer has to do with two facts of nature: (1) the Second Law of Thermodynamics, or the fact that there's an arrow of time in our universe that leads to entropy and the running down of everything, and (2) the logic of evolution, or the fact that natural selection created mortal beings in order to preserve their immortal genes. In answering this question we must also distinguish between direct *proximate* causes and distant *ultimate* causes. *Proximate* causes are more immediate mechanical explanations for why things work the way they do, whereas *ultimate* causes tend toward deeper explanations for why things are a particular way. The *proximate* cause of why fruit tastes sweet, for example, is that taste receptors on the tongue detect the fructose molecules in ripe fruit and send neurochemical signals to the brain that register the sensation of "sweetness." The *ultimate* explanation for why fruit tastes sweet in the first place involves our evolutionary past in which sweet foods such as ripe fruits were at once both rare and nutritious. Natural selection favored individuals for whom rare and nutritious foods were desired over those who had no such taste, and we are the descendants of those for whom fruits tasted sweet. Sex is subject to similar proximate-ultimate causal explanations. *Proximately*, sex feels good because the sex organs are rich in neurons that transduce touch into neurochemical signals that register in areas of the brain associated with pleasure, like the insula and anterior cingulate. *Ultimately*, sex feels good because it is evolution's way of propagating the species and natural selection favored individuals for whom sex was enjoyable. Those for whom sex was unpleasant or neutral were outcompeted by those for whom it was pleasurable because the latter left behind more offspring than the former.

Proximate causes of death are readily apparent and subject to the *availability heuristic*, or the tendency to assign probabilities of potential outcomes based on examples that are immediately available to us, especially those that are vivid, unusual, or emotionally salient.[4] Thus, our initial assessment of what is most likely to kill us—terrorist bombings, shark attacks, earthquakes, hurricanes, lightning strikes, police brutality, killer bees—is whatever happens to be on the evening news at the time we're thinking about it.[5] In fact, these are not the things most likely to kill us. According to the World Health Organization, the top ten killers are ischemic heart disease, stroke, COPD (chronic obstructive pulmonary disease), lower respiratory infection, trachea/bronchus/ lung cancers, HIV/AIDS, diarrheal diseases, diabetes, automobile accidents, and hypertensive heart disease.[6] Other common risks far more likely to take you out before a terrorist or a shark will are cancer (liver, colorectal, breast, skin, prostate, cervical, pancreatic, etc.), poisoning, falls, drowning, fires, injury, drug overdose, and, if you live in the United States, guns—as many Americans die by gun homicides, suicides, and accidents each year as die in automobile accidents (33,636 and 33,804 respectively in 2013).[7]

The *ultimate* cause of death has long had a ready-made answer from theologians and religious believers, who contend that death is simply a transition from one stage to the next. This life is an interim theater from which we exit and receive our divine script for the next act. In the religious worldview, death needs no explanation other than "God wills it" as part of a deific design that will be disclosed once we get to the other side. This may be a satisfying account for some, but it doesn't answer the question of why *physical* life must end, or shape-shift from biological to spiritual, even in a religious worldview. Since God is omnipotent and omnibenevolent, why couldn't he just have created a heaven on earth and skipped the intermediate stage?

For scientists, the ultimate answer to why we age and die begins (and ends) with the Second Law of Thermodynamics, which guarantees that the cosmos is running down and in the long run must come to an end hundreds of billions of years from now. This results in entropy, and it applies to a closed system, which the entire universe is. So yes, ultimately, we—and the earth, the sun, and the universe itself, along with all life-forms therein—must come to an end. But Earth is an open system because of the energy generated by the sun. In principle, as long as there is a source of energy feeding into this open system, life could continue at least another four billion years, at which point the sun will expand outward and envelop the Earth. In the meantime, why can't organisms live indefinitely?

Actually, some appear to do just that. Some organisms do not appear to age (or at least they exhibit *negligible senescence*), such as some tortoises and turtles, sturgeon, rougheye rockfish, and lobsters. Hydras may very well be biologically immortal. In 2016, scientists discovered a Greenland shark that may be the longest-lived vertebrate at 392 ± 120 years, for a range of 272 to 512 years.[8] One specimen is computed to have been born in 1504, sixty years before Shakespeare's birth! Even more extreme are *tardigrades*—water-born, eight-legged microanimals about half a millimeter long found nearly everywhere on Earth and capable of surviving temperature ranges from -458° to +300° F, atmospheric pressures six times greater than the deepest oceans, radiation levels hundreds of times higher than those that would kill humans, no food or water for decades, and even the vacuum of outer space. They are also capable of *cryptobiosis*, a state in which all metabolic processes cease but the organism does not die—a type of suspended animation that may last for thousands of years. If tardigrades can do it, why can't humans? In fact, as we saw in the discussion of the identity problem in chapter 7, you are not the same "stuff" you were at birth because your atoms are recycled and replaced to the point where approximately every decade you are an entirely new person. Why can't that material recycling go on indefinitely, or at least as long as there are atoms to recycle and the energy to drive the process?

To answer these questions we need a precise definition of aging. According to a 2007 paper in the journal *Clinical Interventions in Aging* that reviewed the vast literature on "The Aging Process and Potential Interventions to Extend Life Expectancy," aging "is commonly defined as the accumulation of diverse deleterious changes occurring in cells and tissues with advancing age that are responsible for the increased risk of disease and death." With hundreds of theories of aging on the offing, this plurality of ideas means that we are not close to a consolidation of solutions to slowing or halting the aging process, much less reversing it. Thus, the authors conclude, "the search for a single cause of aging has recently been replaced by the view of aging as an extremely complex, multifactorial process. Therefore, the different theories of aging should not be considered as mutually exclusive, but complementary of others in the explanation of some or all the features of the normal aging process." As for remedies to aging, these scientists are not hopeful: "To date, no convincing evidence showing the administration of existing 'anti-aging' remedies can slow aging or increase longevity in humans is available."[9] Even if we found cures for all the leading proximate causes of death in old age (heart disease, stroke, cancer, etc.), the medical doctor and aging expert Leonard Hayflick has computed that it would add only about fifteen years to human life expectancy.[10] Instead of thinking of aging as a disease for which we may one day find a cure, it is more accurate to describe it as the result of the deterioration of cells and their inability to continue dividing. Why do cells deteriorate and cease dividing?

We still do not know for certain, but the modern Methuselah quest began in 1951 with the Nobel laureate biologist Peter Medawar's now epochal lecture titled "An Unsolved Problem of Biology," in which he contrasted the "wearing out" theory of aging (physics) with the "innate senescence" theory of aging (biology). Medawar opted for the former, employing an analogy with test tubes in his lab in which the glass pipes don't age gradually so much as they break suddenly. Applying his analogy to biology, Medawar turned to anecdotes from trappers who never seem to nab old and senile animals in their steel jaws, and anglers who never report catching senescent fish. In both cases the reason is that such organisms die from accident or predation before they reach old age. (They break before they chip and crack.) Plus, Medawar reasoned, natural selection operates on organisms in their prime reproductive age, so why would there be any reason for evolution to "select" against older members of a species? Without a genetic basis to senescence, aging and death must be the result of the entropy of wearing out. Thus, Medawar defined senescence "as that change of the bodily faculties and sensibilities and energies which accompanies ageing, and which renders the individual progressively more likely to die from accidental causes of random incidence." By this definition, then, "all deaths are in some degree accidental. No death is wholly 'natural'; no one dies merely of the burden of the years."[11]

Not so, says Leonard Hayflick in a 2007 response article titled

"Biological Aging Is No Longer an Unsolved Problem." Hayflick's solution to the problem involves identifying the "common denominator that underlies all modern theories of aging," which is "change in molecular structure and, hence, function." Hayflick's ultimate explanation for aging and death is an "increasing loss of molecular fidelity or increasing molecular disorder," which when you think about it is really just the physical entropy of cells. "The weight of the evidence indicates that genes do not drive the aging process but the general loss of molecular fidelity does." Hayflick explains:

> Unlike any disease, age changes (a) occur in every multicellular animal that reaches a fixed size at reproductive maturity, (b) occur across virtually all species barriers, (c) occur in all members of a species only after the age of reproductive maturation, (d) occur in all animals removed from the wild and protected by humans even when that species probably has not experienced aging for thousands or even millions of years, (e) occur in virtually all animate and inanimate matter, and (f) have the same universal molecular etiology, that is, thermodynamic instability.[12]

That sounds more like physics than biology.

More recently, in 2016 the physicist Peter Hoffman undertook a study of aging after he wrote a book called *Life's Ratchet*, on molecular machinery in cells that prevents them from descending into total entropy. Hoffman was contacted by aging researchers who were curious to know more about how these cellular systems might be maintained indefinitely. They can't, Hoffman explained, because ultimately entropy and the many assaults on cells that accumulate in the course of a lifetime, however long- (or short-) lived, would result in death. "Tinkering with constant risk is helpful, but only to a point. The constant risk is environmental (accidents, infectious disease), but more of the exponentially increasing risk is due to internal wear." Thus, he concludes, "we need to be clear about one thing: We'll never defeat the laws of physics." Although Hoffman titled his article "Physics Makes Aging Inevitable, Not Biology," it seems to me that it is both physics *and* biology that ultimately doom all living organisms. After all, biological systems can be reduced to physical processes, which are ultimately governed by the laws of physics, including the Second Law of Thermodynamics.

So I fail to see the distinction between the physics and biology of aging as anything but arbitrary.

To that end, two of the most oft-cited causes of the physical and biological decay of cells are the *free radical theory of aging* and the related *mitochondrial theory of aging*. Mitochondrial respiration—the generation of energy by converting macronutrients into ATP (adenosine triphosphate) through the use of oxygen in the mitochondria of cells—leaves oxidative free radicals that damage DNA, proteins, and lipids because they have an unpaired electron that seeks to find another electron to pair with from neighboring molecules, thereby damaging them. Breaking these atomic bonds in molecules can cause cancer, as well as lead to the formation of arterial plaques that can result in heart disease and stroke. Antioxidants can help reduce damage from free radicals because they can lose an electron without themselves becoming a free radical, and this has led to endless antioxidant supplements ingested in the form of vitamins A, C, and E. Unfortunately for this otherwise coherent theory, according to a review article in the journal *Free Radical Biological Medicine*, "the current evidence is insufficient to conclude that antioxidant vitamin supplementation materially reduces oxidative damage in humans."[13]

More promising as a causal explanation, albeit not yet for curative treatment, is the *gene regulation theory of aging*, which posits that senescence is the result of changes in gene expression as one grows older, although it is not clear whether the process of genes turning on causes aging itself or simply stops preventing it. It is obvious that one can breed for longevity in animal models, and that in humans one of the best predictors of how long one will live is the longevity of one's biological parents. In other words, longevity runs in the genes. So research is now being conducted on genetic engineering and the use of stem cells, but aging is multicausal across many systems. Plus, broad-based gene manipulation may result in unintended consequences because of a genetic phenomenon called *pleiotropy*, or the production of two or more apparently unrelated phenotypic effects by a single gene (the phenotype is the physical expression of the genotype in interaction with the environment). The selection (or engineering) of one gene with the intention of producing one characteristic may result in the expression of additional unintended characteristics. A famous example is that of the selective breeding for domesticity in silver foxes (*Vulpes vulpes*) by the Russian geneticist Dmitri Belyaev. The normally humanphobic foxes were bred for friendliness toward people, defined by a series of criteria, from the animal allowing itself to be approached, to being hand-fed, to being petted, to actively seeking to establish human contact. In only thirty-five generations (remarkably short on an evolutionary time scale), the researchers were able to produce tail-wagging, hand-licking, pacific foxes. What they also fashioned were foxes with skulls, jaws, and teeth smaller than their wild ancestors, along with floppy ears, curly tails, and striking color patches on their fur, including a star-shaped pattern on the face similar to those found in many breeds of dogs.[14]

The biologist G. C. Williams was the first to link pleiotropy to the evolution of senescence in 1957 in a phenomenon he called *antagonistic pleiotropy*, arguing that traits beneficial to an organism early in its life may be detrimental later in life, such as women's high ovarian steroid levels during peak reproductive age that can lead to breast cancer decades later, or high testosterone in young men that leads to prostate cancer in old age.[15] So genetically engineering longevity, even if it could help push our maximum life-span beyond about 125 years (which it cannot), might result in unintended and possibly antagonistic pleiotropic effects as yet unknown.

The most promising research for breaking through the upper ceiling involves the *telomere theory of aging*, first proposed by the aforementioned Leonard Hayflick, famous for his discovery of the eponymous "Hayflick limit," or the number of times that a normal human cell can divide before cell division stops.[16] The reason has to do with *telomeres*, repeated nucleotide segments at the end of DNA molecules, some of which are lost every time a DNA molecule replicates. When none are left, the cell can no longer divide. There is an association between shortened telomeres and the onset of aging, and researchers have found that the enzyme *telomerase* impedes the shortening process, so perhaps telomerase might delay aging.[17] The overused but instructive analogy is with the plastic ends of shoelaces that become worn until they break off and the lace material unravels. The good news is that there are immortal cells for which this does not happen. The bad news is that they are cancerous cells, so any "cure" for aging involving telomeres will have to avoid also producing cancer.[18] There are some hopeful experiments in which skin cells exposed to an outside source of telomerase stop aging, and in some cases even appear to reverse senescence.[19] Yet other research shows that telomeres cannot be the whole, or even significant, cause of aging, as not all cells that age show telomere shortening, and there are even examples of organisms whose telomeres grow longer with age but they nevertheless experience senescence.[20] To whatever extent telomeres matter for longevity, it is encouraging to note that we may not have to wait for genetic engineering to do something about it, as a 2013 pilot study found that diet (plant-based) and exercise (cardiovascular at least thirty minutes a day, six days a week) increased subjects' telomere length by 10 percent,[21] although with a sample size of only thirty-five men we shouldn't start counting our extra years just yet. Still, that's encouraging, and a 2017 book titled *The Telomere Effect* by the Nobel laureate Elizabeth Blackburn documents how a number of lifestyle changes such as eating well, exercising regularly, and sleeping soundly can improve telomere integrity and possibly even longevity.[22]

If you want to go in whole hog on attenuating aging and delaying death, you might consider adopting Strategies of Engineered Negligible Senescence (SENS), the brainchild of an energetic biomedical gerontologist named Aubrey de Grey, editor of the journal *Rejuvenation Research* and author of the blindly optimistic *Ending Aging*.[23] Aubrey is the tireless promoter of the belief that our generation will be the first to achieve immortality, or at least to live indefinitely, and he's on record claiming that the first human to live a thousand years is alive today.[24] An inheritance has enabled de Grey to build his SENS Research Foundation into a viable institute respectable enough to garner start-up money from Silicon Valley giants like Peter Thiel.[25] If you've seen any television show or documentary film on aging, you've seen the inimitable Aubrey de Grey with his waist-length ponytail and Methuselah-like beard, expounding on life, the universe, and everything in a baritone British accent. I've met Aubrey and shared a beer or two with him (if there is a fountain of youth in de Grey's world, it is a spring that bubbles beer) as he leaned in to bend my ear on the latest shields against the grim reaper's scythe. In brief, if we want to understand aging in order to forestall death, de Grey says we must focus our research efforts on these seven types of cellular damage:

1. Chromosome mutations in nuclear DNA that lead to cancer.
2. Mitochondria mutations in the DNA that can disrupt cellular energy production and progressive cellular degeneration.

3. Intracellular junk (junk inside cells) that results from the breakdown of proteins and other molecules that can build up and lead to atherosclerosis and neurodegenerative diseases such as Alzheimer's.

4. Extracellular junk (junk outside cells, or extracellular aggregates) such as the amyloid plaques entangling neurons in the brains of patients with Alzheimer's.

5. Cellular loss at a rate higher than cellular replacement in youth that leads to general weakness of organs, such as loss of skeletal muscles and heart muscle, loss of neurons leading to Parkinson's, and loss of immune cells that compromises the immune system.

6. Cellular senescence in which cells reach their Hayflick limit and can no longer divide.

7. Extracellular protein crosslinks between cells that cause tissues to lose elasticity and develop problems like arteriosclerosis.[26]

De Grey's SENS Research Foundation has recommendations of what might be done about these assaults on our cells. Whether meeting these seven challenges would result in immortality, or living a thousand years, or even breaking the 125-year ceiling, however, is unknown because we are not even close to accomplishing any of them. Although a 2005 assessment of the program in MIT's *Technology Review* concluded that "SENS does not compel the assent of many knowledgeable scientists; but neither is it demonstrably wrong,"[27] a 2005 study of SENS in the molecular biology journal *EMBO Reports* (the voice of the European Molecular Biology Organization) concluded that none of the therapies recommended by de Grey to counter the SENS seven "has ever been shown to extend the lifespan of any organism, let alone humans."[28] Even de Grey's own SENS Research Foundation admits: "If you want to reverse the damage of aging right now I'm afraid the simple answer is, you can't."[29]

Still, hope springs eternal, and as *Scientific American* reported in a 2016 article intriguingly titled "Is 100 the New 80?" there are some promising preliminary results on the antiaging properties of the diabetes drug metformin, which the FDA approved for clinical trials in 2015.[30] Another 2016 article in *Scientific American*, boldly titled "Aging Is Reversible—at Least in Human Cells and Live Mice," reports on a Salk Institute study in which four mouse genes were turned on that

converted adult cells back into an embryonic-like state, thereby rejuve-
nating damaged muscle cells in a middle-aged mouse. The same was
done with human cells (but not whole human bodies), suggesting that
the epigenetic shift that leads to aging can be reversed by reprogramming specific genes.[31] Unfortunately, some of the treated mice developed tumors and died within a week, so let's not delude ourselves into
believing radical life extension is around the corner.

This reality was well captured in a final pronouncement in *Scientific
American* by S. Jay Olshansky, Leonard Hayflick, and Bruce A. Carnes,
three of the world's leading scientists in the field of aging: "no currently
marketed intervention—none—has yet been proved to slow, stop or
reverse human aging, and some can be downright dangerous."[32] They
note that in addition to the unproven capacity of antioxidants to attenuate the deleterious effects of free radicals on cells, another popular antiaging nostrum called *hormone replacement therapy* may be effective for
some short-term deficiencies such as loss of muscle mass and strength in
older men and postmenopausal women, but the long-term negative side
effects are still unknown and the slowing of the aging process unproven.
Severe calorie restriction appears to work to decelerate aging and
increase maximum life-span in a handful of species such as yeast, fruit
flies, worms, rodents, and fish, but it is not clear it would work on
humans even if you didn't mind spending the rest of your life in a state
of perpetual hunger. As one wag said, "You call that living?" I don't.
"Anyone purporting to offer an antiaging product today is either mistaken or lying," Olshansky, Hayflick, and Carnes conclude. "It is an
inescapable biological reality that once the engine of life switches on,
the body inevitably sows the seeds of its own destruction." Given these
realities, we might as well eat, drink, and be merry . . . and have a beer. A
2016 study by the Mediterranean Neurological Institute in Italy concluded that imbibing a beer or two a day reduces the risk of heart disease
by as much as 25 percent.[33] Bottoms up.

Some of the radical life extension scientists and cryonics proponents
I have met challenge me thus: Wouldn't you like to live to be 200,
500, or 1000 years old? My reply: Sure, of course, but instead of such
lofty goals that are very likely unattainable in my lifetime, if ever, I would
be satisfied with living to be 90 without getting cancer, 100 without
Alzheimer's, 110 without senility, and 120 without being bedridden,

immobile, and insentient. Let's solve those problems first before we worry about what happens at age 200, 500, or 1000.

THE ULTIMATE REASON WE DIE

An ultimate explanation for why we grow old and die comes out of evolutionary theory and was well expressed by the physician Sherwin Nuland: "We die so that the world may continue to live. We have been given the miracle of life because trillions upon trillions of living things have prepared the way for us and then have died—in a sense, for us. We die, in turn, so that others may live."[34]

The technical name for this declaration is the *disposable soma theory of aging*, soma being all the cells in the body except the germ cells, which are disposable after reproduction. It's a Darwinian argument, first proposed by the evolutionary biologist Thomas Kirkwood in 1977.[35] Once a body has passed prime reproductive age (in humans roughly age forty) there's no reason to pour precious resources into it when they can be better invested in offspring. The evolutionary biologist Steven Austad tested the disposable soma theory of aging on two populations of opossums, one on an island with no predators and the other on the mainland with the usual assortment of predators and threats to life and limb. He found that the island opossums reproduced later and aged slower than the mainland opossums.[36]

What about in human populations? The anthropologist Richard Bribiescas recalls witnessing many young girls from the traditional hunter-gatherer Aché people of Paraguay age rapidly once they started having children. Why? Entropy. "The daily grind of activities necessary to care for a family surely contributes to their physical decline." At his lab at Yale University, Bribiescas's research group hypothesized "that women who have more children will exhibit physiological signs of accelerated aging," and they tested this hypothesis on a group of rural postmenopausal Polish women who were part of a long-term study on women's health by the University of Krakow. The researchers found that women with more children had significantly higher levels of oxidative stress compared to those who had fewer children, which is revealing because oxidative stress is one of the key physiological markers of genetic, cellular, and tissue damage associated with aging in all organisms. In the end there's only so much you can do, Bribiescas concludes: "Place an individual into a perfect environment without hazards and with the perfect diet, cognitive stimulation, and every resource that one could identify to maximize life span and you will still eventually end up with a corpse."[37]

Of course, in a species with an extended childhood like ours, parents are needed for many years after birth, and even grandparents can play a vital role in childcare and as repositories of knowledge and wisdom, so the decline is a gradual one lasting decades. But once your children's children are of prime reproductive age, what use are you, really? This is why most organisms grow and flourish throughout childhood and into young adulthood and prime reproductive age, after which the deleterious effects of aging begin to accumulate. Whether the aging process happens by deterioration and the slowing of the maintenance of cells (passive aging) or through programmed cell death (active aging) is not fully understood and remains a theoretical debate among scientists who study the evolution of aging,[38] but it is clear that any ultimate explanation for death must have at bottom an evolutionary framework undergirding it.[39]

The language I've employed here makes it sound as if there's someone directing the evolutionary process, keeping track of investments and payoffs and allocating energy and resources for the good of the species. Not so. There is no intelligent agent running the show from on high, and no goal-directed process operating "for the good of the species" in the future. What Darwin demonstrated is that there is a bottom-up process called natural selection that results in *apparent* design. But the design is a functional by-product of an undirected process. Think about aging and death from the perspective of the genes instead of the body. Genes consist of self-replicating molecules housed inside a cell that contains machinery for energy consumption, maintenance and repair, and other features that keep these molecular structures intact long enough to reproduce. Once such molecular machinery is up and running, the replicating molecules become immortal as long as there is energy to feed the system and an ecosystem in which these processes can take place. Over time these replicating molecules outsurvive nonreplicating molecules by virtue of the very process of replication—those that don't, fail to continue—thus the bodies in which the replicators are housed are survival machines. In this perspective, most famously proposed by the evolutionary biologist Richard Dawkins in his book *The Selfish Gene*, replicators are called genes and the survival machines are called organisms.[40] A survival machine is the gene's way of perpetuating itself into the future indefinitely. Genes that code for proteins that build survival machines that live long enough for them to reproduce win out over genes that do not. From there, selection forces grow weaker for maintaining the robustness of survival machines in old age, and so the life-span of an organism is a balance between the selection pressures that increase reproductive fitness and those that weaken it. The result is mortal bodies and immortal genes.[41] This means that individuals are mortal but species are immortal, as long as they do not go extinct. In our case, the only way to guarantee that this does not happen is to become a multiplanetary species. We are in the process of doing just that.

THE IMMORTAL SPECIES

In 2016, Tesla and SpaceX CEO Elon Musk announced his program for establishing a permanent colony on Mars, starting with transporting one hundred people in eighty days to the red planet, and then repeating the process through reusable rockets until the colony becomes sustainable, perhaps as soon as a century from now. This would make our species multiplanetary, thereby ensuring our survival should something catastrophic happen on our home planet. "Without someone with a real ideological commitment," Musk said in his stirring presentation accompanied by a video enactment, "it didn't seem we were on any trajectory to become a spacefaring civilization."[42] From there it is only a matter of time to island hop from Mars to the moons of Jupiter and Saturn, and eventually work our way out to colonizing exoplanets around other stars. How does this make our species immortal?

There is no known mechanism—short of the end of the universe itself many billions of years from now[43]—to cause the extinction of all planetary and solar systems at once, so as long as our species inhabits multiple planets and moons we can continue indefinitely.[44] In the far future, civilizations may become sufficiently advanced to colonize entire galaxies, genetically engineer new life-forms, terraform planets, and even trigger the birth of stars and new planetary solar systems through massive engineering projects.[45] Civilizations this advanced would have so much knowledge and power as to be essentially omniscient and omnipotent. What would you call such a sentience? If you didn't know the science and technology behind it you would call it God, which is why I have postulated that *any sufficiently advanced extraterrestrial intelligence or far future human is indistinguishable from God.*[46]

It would be too much to say that this form of species immortality satisfies our personal desire to live forever, but it is something well worth working toward, inasmuch as for all we know, we are the only sentient species in the cosmos, so it is incumbent upon us to survive and thrive so that the cosmos may continue to be self-aware. Moreover, given the distances and time scales involved, even if we are the only spacefaring species in the galaxy there is a good chance that each colonized planet will act like a new "founder" population from which a new species evolves, because as the great evolutionary biologist Ernst

Figure 11-1. Per Aspera Ad Astra

From *Finland in the Nineteenth Century*, published in 1894, illustrated by Finnish artists.[47] Translation: *To the stars through struggle.* Sometimes rendered as I present it: *To the stars through audacity,* or *per audacia ad astra.*

Mayr defined it, "a species is a group of actually or potentially inter-breeding natural populations reproductively isolated from other such populations."[48] Different planets, solar systems, and galaxies will act as reproductive isolating mechanisms, and our genus *Homo* will return to a state it last saw tens to hundreds of thousands of years ago when multiple big-brained bipedal apes roamed the planet driven by hunger, lust, and wanderlust out of Africa, out of Asia and Europe, and out of Earth. Thus do we achieve immortality as a species by going to the stars.

Per audacia ad astra.[49]

CHAPTER 12

IMAGINE THERE'S NO HEAVEN

Finding Meaning in a Meaningless Universe

Imagine there's no heaven . . . and at once the sky's the limit.

—Salman Rushdie, the *Guardian*, 1999[1]

The nineteenth-century American poet and novelist Stephen Crane—noted for his realistic depiction in fiction of the psychological complexities of life, as in his classic Civil War novel *The Red Badge of Courage*—expressed the problem of finding meaning through this five-line, twenty-four-word masterpiece of brevity, a humbling of humanity before the vastness of the cosmos:

> *A man said to the universe:*
> *"Sir, I exist!"*
> *"However," replied the universe,*
> *"The fact has not created in me*
> *A sense of obligation."*[2]

Indeed, the universe is under no obligation to acknowledge our existence, much less give it meaning. That obligation is vouchsafed to us and is more readily available with a deeper understanding of our place in time and space. In this terminus to our journey I would like to pull back to look at the bigger picture across both time and space and consider how mortal beings can find meaning in an apparently meaningless universe.

Time. A year represents a sliver of time that amounts to a mere one one-hundred-fifty-thousandth of the life-span of our species, one ten-thousandth of the epoch of our civilization, one five-hundredth of the Age of Science, and one one-hundredth of the Age of Einstein, who discovered that space and time are indivisible. Further, our species is but one among hundreds of millions—perhaps billions—of species that evolved over the course of 3.5 billion years on Earth, itself around 4.6 billion years old and accounting for only a third of the age of the 13.8-billion-year-old universe. If you lived for a century, that would amount to a mere 0.0000000073 percent of the cosmic life-span.

Space. Earth is but one tiny planet among many orbiting an ordinary star, itself one of hundreds of billions of stars in the Milky Way galaxy, the vast majority of which have many planets forming solar systems quite possibly teeming with life. Our galaxy is just one among a cluster of galaxies not so different from millions of other galaxy clusters, whirling away from one another in an accelerating expanding cosmic bubble universe that could very possibly be one among a near infinite number of expanding bubble universes in an unimaginably vast multiverse. Figure 12-1 captures the sense of awe invoked by this Ultra Deep Field image from the Hubble Space Telescope taken in 2014 showing hundreds of galaxies 5 to 10 billion light-years away, meaning that the light from them left before the Earth was even formed.

Is it really possible that our minutely narrow fraction of a percent of that 13.8-billion-year cosmic history is the reason for the universe's existence? Is it really possible that this entire cosmological multiverse was designed and exists for one tiny subgroup of a single species on one planet in a lone galaxy in that solitary bubble universe? To answer these questions in the affirmative would be hubris enough to make even a Greek god blush.

But with this cosmic perspective, where shall we seek meaning and purpose? It begins with a deeper understanding of spirituality and awe.

IN THE AWE

Awe is the wonderment that comes from being humbled before something grander than oneself. Many people think of this awe-inspiring thing as God and call it spirituality, but a great many others find awe in the

Figure 12-1. Hubble Image of Galaxies

In 2014 the Hubble Space Telescope captured this Ultra Deep Field image show-ing hundreds of galaxies 5 to 10 billion light-years away, meaning that the light captured in the camera left those galaxies before the creation of the Earth. Source: NASA, ESA, H. Teplitz and M. Rafelski (IPAC/Caltech), A. Koekemoer (STScI), R. Windhorst (ASU), Z. Levay (STScI).

wonder of the cosmos itself. This was beautifully articulated one Sun-day in 2013 by the long-distance swimmer Diana Nyad when she appeared on Oprah Winfrey's television show *Super Soul Sunday*.[3]

My interest was piqued because I met Diana back in 1982 on the eve of the first three-thousand-mile nonstop transcontinental bicycle Race Across America. She was covering the race for ABC's *Wide World of Sports*. I was riding in it. I knew of her 28-mile swim around Man-hattan, her 102-mile open-ocean swim from North Bimini in the

Bahamas to Juno Beach, Florida, and her unsuccessful attempt to swim from Cuba to Florida. At a prerace dinner in Los Angeles, with my nerves in my throat about undertaking my first transcontinental crossing, I asked her what it was like being a long-distance swimmer and what kept her going through hardship and failure. I don't remember her exact words, but the single-minded intensity and strength of will that came through in her presence inspired me over the next ten days to make it to New York. In 2013, Nyad finally completed her swim from Cuba to Florida on her fourth try, in 52 hours, 54 minutes. It is the longest unaided open-ocean swim in history, and she did it at age sixty-four.

It was the triumph of Nyad's will over age that Oprah wanted to suss out for her audience. Was it something spiritual, a higher power perhaps? No. "I'm an atheist," Nyad explained. Oprah responded quizzically: "But you're in the awe." Puzzled herself, Nyad responded: "I don't understand why anybody would find a contradiction in that. I can stand at the beach's edge with the most devout Christian, Jew, Buddhist—go on down the line—and weep with the beauty of this universe and be moved by all of humanity, all the billions of people who have lived before us, who have loved and hurt and suffered. So to me, my definition of God is humanity and is the love of humanity." What Oprah said next inflamed atheists: "Well, I don't call you an atheist then. I think if you believe in the awe and the wonder and the mystery, then that is what God is." This is the soft bigotry of those who cannot conceive of how someone can be in awe without believing in supernatural sources of wonder. Why would anyone think that?

A partial answer may be found in a 2013 study by the psychologists Piercarlo Valdesolo and Jesse Graham titled "Awe, Uncertainty, and Agency Detection."[4] Previous research had found that "awe" is associated with "perceived vastness" (as in the night sky or an open ocean) and that "awe-prone" individuals tend to be more comfortable with uncertainty and to be less likely to need cognitive closure in an explanation (of any kind). They "are more comfortable revising existing mental schemas to assimilate novel information," the authors wrote in their paper. For people who are not awe-prone, Valdesolo told me in an email, "we hypothesized that the uncertainty experienced by the immediate feeling of the emotion would be aversive (since they are probably not the kinds of people who feel it all the time). This was rooted in theoretical work that argued that awe is elicited when we have trouble making sense of the event we are witnessing, and this failure to assimilate information into existing mental structures should lead to negative states like confusion and disorientation."[5]

To reduce the anxiety of awe-inspiring experiences, non-awe-prone people engage in a process I call *agenticity*, or the tendency to believe that the world is inhabited and controlled by invisible and intentional agents. Applied broadly, I have argued that this is the basis for belief in souls, spirits, ghosts, gods, demons, angels, aliens, and all manner of invisible agents with power and intention to control our lives. When combined with our propensity to find meaningful patterns in both meaningful and meaningless noise—what I have called *patternicity*—the two form the cognitive basis of shamanism, paganism, animism, polytheism, monotheism, and all modes of Old and New Age spiritualisms.[6]

To test this hypothesis, Valdesolo and Graham showed subjects one of three video clips: (1) an awe-inspiring scene from the BBC's television series *Planet Earth*, (2) an emotionally neutral news interview by the late investigative journalist Mike Wallace, or (3) a comedy clip from the BBC's *Walk on the Wild Side*. Subjects then took a survey that measured their belief in God, belief in "supernatural control" (by which the psychologists mean "belief that the universe is controlled by God or supernatural forces, such as karma," on a 1 to 10 scale from "tremendously doubtful" to "extremely likely"), and a measure of their feeling of "awe" while watching the video clip. For this last measure subjects completed an eight-item questionnaire with such questions as "To what extent did you experience awe while watching the video clip?" on a 1 to 7 scale from "not at all" to "extremely."

Unsurprisingly, subjects experienced more awe while watching *Planet Earth* than the comedy clip or the Mike Wallace interview (although, personally, I always found Wallace to be awe-inspiring when he was grilling some corrupt politician or conniving CEO), but they also showed greater belief in both God and supernatural control in the awe condition. Valdesolo and Graham concluded: "The present results suggest that in the moment of awe, some of the fear and trembling can be mitigated by perceiving an author's hand in the experience."[7]

This brings me back to Diana Nyad and those of us who find our spirituality in the awe of the natural world without a need for supernatural agenticity. Instead of fear and trembling, we feel wonder and gratitude in discovering that the author's hand is nature's laws and nothing more . . . but nothing less. In her memoir, *Find a Way*, in a chapter entitled "Atheist in Awe," Nyad recalls a trip to the Amazon in which her hosts' puppy went missing for five days. Nyad was especially sympathetic because she had a dog at home who died during this trip, leaving her feeling powerless to help and distraught at the loss. "I was helpless. I wasn't there to console her," she writes. Later that night Nyad couldn't sleep, so she went out and sat in an open field to write in her journal about her beloved canine, Moses. All of a sudden, her hosts' missing puppy appeared out of the darkness and crawled into her lap. "I'm quite sure the interpretation could easily be made that this little puppy was some kind of sign from the universe, Moses's spirit reincarnated for me to hold close," she reflected on the human propensity to believe that everything happens for a reason. "To me, it was a sad passing followed by a happy coincidence, a cue to embrace the chaos, in itself a paradox of joy and sorrow, life and death existing in the same moment, neither canceling the other out. Just part of the wondrous, random world, not something meant to be."[8]

FINDING PURPOSE IN THE COSMOS

Science reveals that the pattern of information that represents your body and brain as coded in your genome and connectome—including and especially your thoughts and memories, and particularly your point of view—is your soul, and since there is no evidence that this pattern exists separately from your body and brain, or that it continues on after your death, your physical body and your soul are one. The impossibility of duplicating your soul—not just your genome and your connectome but those of everyone who ever lived, along with all the historical events and forces that shaped your life and that of your ancestors, culture, and all of history—means that each one of us is unique in the cosmos. There is no one who is—or ever could be—like anyone else. From this reality, discovered and described as it has been by science, we may derive meaning. How?

Through recognition of our uniqueness, through our gratitude for having the chance to live, through the love of others and others' love for us, and through engagement with the world with courage and integrity. Although immortality has not yet been vouchsafed to us in this universe, we live on nonetheless through our genes and our families, our loves and our friends, our work and our engagement with others, our participation in politics, the economy, society, and culture, and our contributions—however modest—to making the world a little bit better today than it was yesterday. Protopian progress is real and meaningful, and we can all make a mark, however small.

Progress and purpose are intertwined motives from which we may derive deeper meaning in the cosmos, starting with the laws of nature. Such "laws" are the linguistic and mathematical descriptions we humans give to naturally occurring repeating phenomena. In this sense, there are no laws of nature "out there" floating about in the ether autonomous and independent of nature itself. Stars, for example, convert hydrogen into helium in a well-defined manner dependent on temperature and pressure. We can write out mathematical equations and linguistic descriptions that explain and predict how this process happens, how fast, how much, and so forth, but there are no "laws" inside stars; there is just star stuff doing what star stuff must do under those conditions.

And it's all star stuff.

Everything on Earth—the land, the sea, and every living thing on and in it, from basic bacteria to big brains—is made of atoms, whose genesis was the interior of ancient stars whose lives ended in breathtaking outbursts of supernova explosions that created those new and more complex atoms and dispersed them into space where they re-formed into new stars and planets, many of which may contain life, and one of which even houses sentient beings capable of understanding this very process of creation. This revelation was first shared with the world in 1926 by the astronomer Harlow Shapley, through a popular lecture series broadcast on Boston's WEEI radio:

> We are therefore made out of star stuff . . . we feed upon sunbeams, we are kept warm by the radiation of the sun and we are made out of the same materials that constitute the stars.[9]

Stars died so that we may live.

In this sense nature is not purposeless, as so many people seem to think it is, leading them to invoke a supernatural entity outside nature to grant us purpose. No such agent is needed because purpose is built into the cosmos and the laws of nature. The purpose of stars is to convert hydrogen into helium and generate light and heat. This is their "destiny," their cosmic purpose. And this is true for all things in the cosmos, and here on Earth. The "purpose" of mountains is to grow in height as a result of geological forces like plate tectonics, and to shrink in size as a result of erosive forces like weather. The "purpose" of rivers is to find the lowest resting place of water under the force of gravity, and as a result they cut through solid rock and course across entire continents. The "purpose" of life is to survive, reproduce, and flourish, and it has been fulfilling its destiny for 3.5 billion years in an unbroken chain from the Precambrian to today and encompassing all forms of life of which we know.

The difference between mountains, rivers, and living organisms is that life survives, reproduces, and flourishes in the teeth of entropy. This makes the Second Law of Thermodynamics the First Law of Life. As the evolutionary psychologists Leda Cosmides, John Tooby, and Clark Harrett noted in their paper on the purpose of evolution: "The most basic lesson is that natural selection is the only known natural process that pushes populations of organisms thermodynamically uphill into higher degrees of functional order, or even offset the inevitable increase in disorder that would otherwise take place."[10] This "extropy" occurs only in an open system with an energy source, such as our planet with the sun providing the energy that temporarily reverses entropy, and with replicating molecules like RNA and DNA that enable living organisms to send near-duplicates out into the world that provide fodder for natural selection. Once this system is up and running, evolution can move away from the left wall of minimum order and simplicity and toward the right wall of maximum order and complexity. If you do nothing, entropy will take its course and you will move toward a higher state of disorder (ultimately causing your demise). So your most basic purpose in life is to combat entropy by doing something extropic—expending energy to survive, reproduce, and flourish. In this sense, evolution granted us a purpose-driven life by dint of the laws of nature. Purpose is in our nature, and in our case we have brains big enough to contemplate it and language sophisticated enough to talk about it, so our species alone (as far as we know) is capable of understanding and defining what it means to have a purpose-driven life.

Most people think of purpose as being defined by some outside transcendent source that validates all that we do, but what science tells us is
that there is no Archimedean point from which we can lever into our
lives some external purpose beyond nature. The Persian astronomer-poet
Omar Khayyám rhymed it elegantly in this quatrain from *The Rubaiyat*:

> *I sent my Soul through the Invisible,*
>   *Some letter of that After-life to spell:*
> *And by and by my Soul return'd to me,*
>   *And answer'd "I Myself am Heav'n and Hell."*[11]

Heaven and hell are within us, not above and below us.

At St. Paul's Cathedral in London, built by the great architect and
polymath Sir Christopher Wren, who designed so many of London's classic buildings, his epitaph reads: *Si monumentum requires, circumspice*—
"If you seek a monument, look around you." To paraphrase the
inscription in a message for what we may conclude from the human
quest for immortality and perfectibility:

> *Si requires caelo, circumspice.*
> If you seek a heaven, look around you.

Heaven is not "up there" in some celestial empyrean; it is all around
us. We create our own purpose, and we do this by fulfilling our nature,
by living in accord with our essence, by being true to ourselves, as
Shakespeare counseled:

> *This above all: to thine own self be true,*
> *And it must follow, as the night the day,*
> *Thou canst not then be false to any man.*[12]

To thine own self be true. What does this mean? Individually, each
answer must be deeply personal and uniquely our own, but collectively
it begins with the Law of Identity: *A is A*. Being true to yourself means
recognizing and acknowledging that *A is A*, that existence exists, reality
is real, and that you are you and not someone else. To try to be something that you are not, or to pretend to be someone else, is a violation
of the Law of Identity: *A cannot be non-A*.

A is A means discovering who you are, your temperament and personality, your intelligence and abilities, your needs and wants, your loves and interests, what you believe and stand for, where you want to go and how you want to get there, and what matters most to you. *Thine own self* is your *A*, which cannot also be *non-A*. The attempt to make *A* into *non-A* has caused countless problems, failures, and heartaches in people's lives. People who befriend those who are not in harmony with their interests are not being true to themselves. People who love those who do not love them back in the same way are not being true to themselves. People who work in jobs that they hate are not being true to themselves or to those around them. People who seek self-esteem through the success of others are in violation of the true nature and cause of self-esteem: accomplishments through effort. *A* can never be *non-A*, no matter how hard anyone tries to make it so. The Law of Identity is another First Law of Life and is as ironclad as the Second Law of Thermodynamics.

Purpose is personal, and there are many ways to satisfy this deep-seated need. But science informs us that there are some tried-and-true methods by which we can bootstrap ourselves toward purpose by pursuing goals that consistently lead to a greater sense of meaning:

1. *Love and family.* The bonding and attachment to other people increases one's circle of sentiments and a corresponding sense of purpose to care about others as much as, if not more than, oneself.

2. *Meaningful work and career.* Having a passion for work and a long-term career gives most people a drive to achieve goals beyond the needs of themselves and their immediate family that lifts all of us to a higher plane, and society toward greater prosperity and meaning. Having a reason to get up and around in the morning and having a place to go where one is needed is a lasting purposeful activity.

3. *Social and political involvement.* We are not isolated individuals but autonomous social animals with deep feelings for our fellow sentient beings and a drive to participate in the process of determining how best we should live together, for the benefit of ourselves, our families, our communities, and our societies. This means not just voting, for example, but being actively engaged in

the political process; it is not just a matter of joining a club or society, but caring about its goals and the actions of the other members working toward the same goals.

4. *Transcendency and spirituality.* Possibly unique to our species is the capacity for aesthetic appreciation, spiritual reflection, and transcendent contemplation through a variety of expressions such as art, music, dance, exercise, sports, meditation, prayer, quiet contemplation, and religious reverie, connecting us to that which is outside ourselves, and generating a sense of awe at the vastness of humanity, nature, the world, or the cosmos.

5. *Challenges and goals.* Most of us need tests and trials and things at which to aim, both ordinary, such as the physical challenge of sports and recreation and the mental challenge of games and intellectual pursuits, and extraordinary, such as striving for abstract principles such as truth, justice, and freedom and struggling through the obstacles in the way of realizing them.

Research shows that seeking happiness by itself is not enough for most of us. We want meaning in our lives, a sense of purpose that derives from something more than pleasure seeking or pain avoidance. Although happiness and meaning overlap, they are not the same. The social psychologist Roy Baumeister and his colleagues published a study in the *Journal of Positive Psychology* that reported about half the variation in meaningfulness in their sample was explained by happiness and vice versa, but the rest of the variation was dependent upon five areas where happiness and meaning differed.[13]

1. *Wants and Needs.* Satisfying one's desires may lead to greater happiness, but it appears to have little or nothing to do with meaning and purpose. Flying first class on a transcontinental flight may put me in a better mood when I deplane on the other side of the country, but it doesn't make my life any more meaningful. For that I need to actually do something when I get there, and that activity may or may not make me any happier. Good health is another example: it makes people happier to be healthier, but it doesn't contribute anything to meaning. Likewise money: it's better to be rich than it is to be poor ("if only for financial reasons," as Woody Allen joked), but the money alone won't grant you a purposeful life. Money may serve as a barometer of how

your work and investments are progressing (or regressing), and it may make life more comfortable and easier, but comfort and ease are orthogonal to purpose.

2. *Time Frames and Actions*. Happiness is a present state of mind whereas meaning transcends the here and now, reaching back to the past and forward into the future. Thinking a lot about the past and future does not make people more or less happy, but it does lead them to feel more meaning, as if what they have done, are doing, or will do makes a difference. Thinking about only the here and now can make people more or less happy (depending on how the here and now is going) but seems to have no effect on meaningfulness. Although happiness is partially heritable (about half of the variance among people in measures of happiness is accounted for by their genes), it appears that meaning is almost entirely environmental—what you do more than who you are.

3. *Family and Friends*. Baumeister and his team found that both happiness and meaning are linked to other people, although in different directions: *caring for others* increases meaning, whereas having *others care for you* increases happiness. Caring for children or aging parents, for example, may not make one immediately happy, but it does bring a sense of purpose. I was a caretaker for my mother and my stepfather, which I found to be a taxing experience both physically and emotionally. It brought me no happiness but it did produce deep meaning in caring for those who cared for me. By contrast, having someone care for you when you are in need can make you happier. The researchers had their subjects rate themselves as "givers" or "takers" and found that the former reported leading more meaningful lives but being less happy compared to the latter. And the type of social interaction matters. Just hanging out with friends enjoying a meal or activity may increase happiness but it probably does nothing for meaning, whereas spending time with family members may not lead to higher happiness but there may be long-term effects on meaning, higher or lower depending on the state of the relationship—arguing with family, for example, was associated with higher meaning but lower happiness.

4. *Struggles and Problems*. Having good things happen to you leads to both happiness and meaning, while having bad things happen to you reduces happiness but may lead to increased meaning, depending on the

outcome. Purely happy people may lead a stress-free life, but they may not have a meaningful existence, whereas those who seem unhappy or stressed out may in fact be experiencing purpose and fulfillment. Retirement, for example, typically leads to an increase in happiness but a decrease in meaningfulness, until the person finds something goal-directed and challenging to do. In a personal example, writing this book gave me a great sense of meaning and purpose, but it rarely made me happy, and then only when I finished a section, chapter, or the entire manuscript. It's hard to write a book, and I'm always trying to improve as a writer, so I push myself hard to be original and creative and to do so in a literary style that engages other minds, which can be stressful. As my wife will attest, I'm often gruff and grumpy when I'm so focused, purpose driven but not pleasure seeking.

5. *Self and Personal Identity.* Actions that are expressions of your identity and sense of self are most directly related to meaning but least related to happiness. In another personal example, I have been a serious cyclist for my entire adult life; it is how, in part, I define myself and what and who I identify with, but I am rarely happy doing it. As in most sports, when you push yourself hard there is a lot of suffering in cycling. A "sufferfest" is how we often describe races and intense training rides. It gives me great meaning (accompanied by a lot of endorphins) to do a long, hard ride, but for most cyclists, "happiness" isn't even in our vocabulary when we're cycling.

In an essay summarizing the results of this research, Baumeister captured what I am trying to convey about the purpose of life, the laws of nature, and the cosmos as it relates to finding meaning, particularly in the context of our search for immortality, the afterlife, and utopia:

> Meaning is a powerful tool in human life. To understand what that tool is used for, it helps to appreciate something else about life as a process of ongoing change. A living thing might always be in flux, but life cannot be at peace with endless change. Living things yearn for stability, seeking to establish harmonious relationships with their environment. They want to know how to get food, water, shelter and the like. They find or create places where they can rest and be safe . . . Life, in other words, is change accompanied by a constant striving to slow or stop the process of change, which leads ultimately to death. If only change could stop, especially at some perfect point: that was the theme of the

profound story of Faust's bet with the devil. Faust lost his soul because he could not resist the wish that a wonderful moment would last forever. Such dreams are futile. Life cannot stop changing until it ends.[14]

That a meaningful, purposeful life comes from struggle and challenge against the vicissitudes of nature more than it does a homeostatic balance of extropic pushback against entropy reinforces the point that the Second Law of Thermodynamics is the First Law of Life. We must act in the world. The thermostat is always being adjusted, balance sought but never achieved. There is no Faustian bargain to be made in life. We may strive for immortality while never reaching it, as we may seek utopian bliss while never finding it, for it is the striving and the seeking that matter, not the attainment of the unattainable. We are volitional beings, so the choice to act is ours, and our sense of purpose is defined by reaching for the upper limits of our natural abilities and learned skills, and by facing challenges with courage and conviction. This idea was expressed poetically by William Ernest Henley in his 1920 poem "Invictus," particularly poignant in that he wrote it when he was terminally ill:

Out of the night that covers me,
Black as the pit from pole to pole,
I thank whatever gods may be
For my unconquerable soul.

In the fell clutch of circumstance
I have not winced nor cried aloud.
Under the bludgeonings of chance
My head is bloody, but unbowed.

Beyond this place of wrath and tears
Looms but the Horror of the shade,
And yet the menace of the years
Finds and shall find me unafraid.

It matters not how strait the gate,
How charged with punishments the scroll,
I am the master of my fate:
I am the captain of my soul.

LOVE, LIFE, AND DEATH

To return to where we began our journey, in the first chapter we considered one of the central themes among the final statements of the Texas death row inmates as they faced their own mortality. Lying supine crucifix-like on the gurney awaiting lethal injection, the last thing the majority of those about to die experienced was love. Facing one's end brings this reality into focus as no intellectual exercise can. The psychiatrist Viktor Frankl came to much the same conclusion in his now classic 1946 work *Man's Search for Meaning*, in which he reflected on how one can find meaning when facing death—in his case, at Auschwitz:

> A thought transfixed me: for the first time in my life I saw the truth as it is set into song by so many poets, proclaimed as the final wisdom by so many thinkers. The truth—that love is the ultimate and the highest goal to which man can aspire. Then I grasped the meaning of the greatest secret that human poetry and human thought and belief have to impart: *The salvation of man is through love and in love.* I understood how a man who has nothing left in this world still may know bliss, be it only for a brief moment, in the contemplation of his beloved. In a position of utter desolation, when man cannot express himself in positive action, when his only achievement may consist in enduring his sufferings in the right way—an honorable way—in such a position man can, through loving contemplation of the image he carries of his beloved, achieve fulfillment.[15]

This positive side to the existential considerations of facing death was documented by the psychologist Kenneth Vail and his colleagues in a series of papers on the salubrious effects of *mortality salience* (MS), or reminding people of their mortality.[16] For example, MS improves physical health, encourages life-goal reflection, motivates prosocial behavior, fosters loving relationships, inspires community involvement, stimulates concerns for the environment, and even supports intergroup peace building. People who are reminded that they are mortal, for example, are more likely to wear a seat belt, quit smoking, use sunscreen, get medical checkups, and exercise. Experiments find that MS bolsters efforts to achieve personal goals, both physical (e.g., basketball performance) and mental (e.g., reading comprehension), and motivates people to improve their personal relationships, both socially (e.g., with friends and co-workers) and romantically (e.g., with dates or spouses). MS also leads people to devalue extrinsic goals like fame and wealth and to emphasize intrinsic values like personal relationships. Divorce rates following the 1995 Oklahoma City bombing, for example, plummeted in the surrounding counties, suggesting that death awareness may lead people to make a stronger commitment to their marriages and families. There's even a term for this: *post-traumatic growth*. PTG may, in part, help explain the positive response many people have to near death experiences, in which they report having a greater appreciation for life, less interest in material goods and social status, and above all a deeper love for friends and family. Most revealing for the only proven form of immortality we know of—propagating our genes into the next generation—a number of studies found that reminding people of their mortality intensified their intent to have and care for children.

There are wider implications for society of facing death in a positive fashion. Being reminded of mortality, for example, also enhances social group identity and community engagement and encourages people to treat one another more fairly and compassionately (although there is a potential downside inasmuch as such group cohesiveness can lead to tribalism, xenophobia, and even violence against outsiders). Studies show that MS, for example, increases donations to charities, contributions to community youth groups, aid to services for the elderly or disabled, and acting in a more environmentally conscious manner among those who value being green. It has even been shown to make people feel more inclusive of others not normally in our social groups.

Participants reminded of global warming, for example, were more supportive of international peacemaking, in the sense that a threat to all of us reduces the concerns about the differences between us. Most remarkably, in a study conducted among Arabs in Israel during that country's invasion of Gaza in January 2009, "mortality salience increased support for peaceful coexistence with Israeli Jews among those who imagined global warming and had high perceptions of common humanity," Vail and Juhl write. "Together, this work illustrates that situations that foster more inclusive superordinate group identifications can create conditions where existential motivation can lead to a more inclusive

treatment of individuals who might otherwise be out-group members."[17] Perhaps the long-delayed two-state solution needs something
existentially threatening to both Arabs and Israelis to at long last bring
it about.[18]

———

FACING DEATH—AND life—with courage, awareness, and honesty can
bring out the best in us and focus our minds on what matters most:
gratitude and love. Gratitude for a chance at life, given the biological
reality that those hundred billion people who lived before us were, in
fact, only a tiny fraction of the many trillions of people who could have
been born but were not. The chance encounter of sperm and egg that
led to each of us could just as well have produced someone else, and
you would never know it because there would be no you to know. Once
born, we are each unique, a concatenation of genes and brains with
thoughts, feelings, memories, histories, and points of view that can
never be duplicated, here or in the hereafter. Our sentience—yours, mine,
everyone's—is ours alone and like no other anywhere in the cosmos. We
are given this one chance to live, some four score trips around the sun, a
brief but glorious moment in the cosmic drama unfolding on this provisional proscenium. Given all we know about the universe and the laws
of nature, that is the most any of us can reasonably hope for. Fortu-
nately, it is enough. It is the soul of life. It is heaven on earth.

NOTES

PROLOGUE

1. Quoted in: Warren, James (ed.). 2004. *Facing Death: Epicurus and His Critics*. Oxford: Clarendon Press, 19.
2. I arrived at these figures from data provided by the Population Reference Bureau of the U.S. Census in Washington, D.C.: http://bit.ly/19LQkBa. Carl Haub is the demographer who computed the numbers through 2011 in his fascinating article "How Many People Have Ever Lived on Earth?" I updated the estimates through 2017. As Haub makes clear, the estimation of how many people have ever lived is necessarily a rough approximation, since accurate population census data were not available until recently. Still, the numbers are sufficiently accurate to make my point.
3. For current population estimates and an ongoing world population clock see: http://bit.ly/Ism9T6
4. The ratio of the dead to the living will fall from my calculated figure of 14.4 to 1 to about 11 to 1 by 2050 according to the website FiveThirtyEight: http://53eig.ht/1khiL04. The first person I know of to make such calculations was Arthur C. Clarke on the opening page of his book *2001: A Space Odyssey*.
5. The claim that some people have gone to the other side (usually heaven) and returned to report what they saw does not meet the high standards required by science in order to be considered valid. I will address these claims in a later chapter in this book and have dealt with these claims in detail in two of my previous books: Shermer, Michael. 1997. *Why People Believe Weird Things*. New York: Times Books; Shermer, Michael. 2011. *The Believing Brain*. New York: Henry Holt.

6. Global Health Observatory Data Repository. 2011. World Health Organization. http://bit.ly/1ediO26. The *maximum life potential* (the age at death of the longest-lived member of the species) has not changed and remains at 120 years. *Life-span* (the age at which the average person would die if there were no premature deaths from accidents or disease) has also not changed and remains at around 85 to 95 years. But *life expectancy* (the age at which the average individual would die when accidents and disease have been taken into consideration) has skyrocketed upward from 47 years in 1900 in the United States to 78.9 for all Americans born in 2010, and 85.8 for Asian American women.

7. Goldenberg, David. 2015. "Why the Oldest Person in the World Keeps Dying." FiveThirtyEight, May 26. http://53eig.ht/1TD6D42

8. Dylan Thomas's poem "Do not go gentle into that good night" can be found here: http://bit.ly/1kzvZ8J. John Donne's sonnet "Death Be Not Proud" can be found here: http://bit.ly/1gLqjyO

9. http://bit.ly/1ObJPWT

10. Greeley, Andrew M., and Michael Hout. 1999. "Americans' Increasing Belief in Life After Death: Religious Competition and Acculturation." *American Sociological Review*, 64(6): 813–835.

11. http://pewrsr.ch/1Za3kJR

12. http://bit.ly/1DMBgwJ

13. Gallup, George, Jr., and William Proctor. 1982. *Adventures in Immortality*. New York: McGraw-Hill.

14. Dugatkin, Lee Alan, and Lyndmila Trut. 2017. *How to Tame a Fox (and Build a Dog): Visionary Scientists and a Siberian Tale of Jump-Started Evolution*. Chicago: University of Chicago Press, 229.

15. http://reut.rs/1OBcZm0. Countries surveyed were Argentina, Australia, Belgium, Brazil, Canada, China, France, Germany, Great Britain, Hungary, India, Indonesia, Italy, Japan, Mexico, Poland, Russia, Saudi Arabia, South Africa, South Korea, Spain, Sweden, Turkey, and the United States.

16. http://bit.ly/1rV3m8Z

17. Quoted in: Bulos, Nabih. 2016. "Saudi Cleric Urges Rebels to Fight, Die." *Los Angeles Times*, August 8, A3. http://lat.ms/2aNVbpY

18. Segal, Alan. 2004. *Life After Death: A History of the Afterlife in Western Religion*. New York: Random House, 659.

19. The reference is to the famous line from the classic 1962 Western film *The Man Who Shot Liberty Valance*: "This is the West, sir. When the legend becomes fact, print the legend." http://bit.ly/2aqaY0s

20. In: Rubin, Barry, and Judith Colp Rubin, eds. 2002. *Anti-American Terrorism and the Middle East: A Documentary Reader*. New York: Oxford University Press, 237.

21. http://bit.ly/2av2S1L

22. See for example: Schwarz, Jon. 2017. "Why Do So Many Americans Fear Muslims? Decades of Denial about America's Role in the World." *Intercept*, February 18. http://bit.ly/2m858S8

CHAPTER 1

1. Fables, Robert, trans. 1984. *Sophocles: The Three Theban Plays: Antigone; Oedipus the King; Oedipus at Colonus.* New York: Penguin.
2. Sartre, Sean Paul. 1948. *Existentialism and Humanism.* Paris: Editions Nagel.
3. Quoted in: Teodorescu, Adriana, ed. 2015. *Death Representations in Literature: Forms and Theories.* Newcastle upon Tyne: Cambridge Scholars Publishing, 12.
4. Freud, Sigmund. 1915. *Reflections on War and Death.* Online reprint: http://bit.ly/1WISuJ1
5. Becker, Ernest. 1973. *The Denial of Death.* New York: Free Press, 26.
6. Cave, Stephen. 2012. *Immortality: The Quest to Live Forever and How It Drives Civilization.* New York: Crown, 19.
7. Allen, Woody. 1975. *On Being Funny: Woody Allen and Comedy.* New York: Charterhouse.
8. Solomon, Sheldon, Jeff Greenberg, and Tom Pyszczynski. 2015. *The Worm at the Core: On the Role of Death in Life.* New York: Random House.
9. James, William. 1902. *The Varieties of Religious Experience.* New York: Longman, 140; available online: http://ntrda.me/1SV5jhf
10. Quoted in: Horgan, John. 2016. Interview with Sheldon Solomon. *Scientific American*, June 6. http://bit.ly/1XydaEg
11. Solomon, S., J. Greenberg, J. Schimel, J. Arndt, and T. Pyszczynski. 2003. "Human Awareness of Mortality and the Evolution of Culture." In: *The Psychological Foundations of Culture.* Mahwah, NJ: Lawrence Erlbaum Associates, 15–40.
12. Pinker, Steven. 2011. *The Better Angels of Our Nature: Why Violence Has Declined.* New York: Viking, 648.
13. Solomon et al. 2003.
14. Guthrie, R. Dale. 2005. *The Nature of Paleolithic Art.* Chicago: University of Chicago Press.
15. Liebenberg, Louis. 2013. "Tracking Science: The Origin of Scientific Thinking in Our Paleolithic Ancestors." *Skeptic* 18(3).
16. Pinker, Steven. 1997. *How the Mind Works.* New York: W. W. Norton, 189.
17. Buss, David M. 1997. "Human Social Motivation in Evolutionary Perspective: Grounding Terror Management Theory." *Psychological Inquiry* 8(1): 22–26.
18. Miller, Geoffrey. 2000. *The Mating Mind: How Sexual Choice Shaped Human Nature.* New York: Doubleday.

19. Personal correspondence, March 13, 2016.
20. Results were published in: Shermer, Michael. 2000. *How We Believe: Science, Skepticism, and the Search for God*. New York: Henry Holt.
21. Personal correspondence, June 27, 2016.
22. Feynman, Richard, and Ralph Leighton. 1985. *Surely You're Joking, Mr. Feynman*. New York: W. W. Norton; Feynman, Richard. 1988. *What Do You Care What Other People Think?* New York: W. W. Norton; Feynman, Michelle, ed. 2015. *The Quotable Feynman*. Princeton, NJ: Princeton University Press.
23. Gleick, James. 1992. *Genius: The Life and Science of Richard Feynman*. New York: Pantheon, 438.
24. Hitchens, Christopher. 2012. *Mortality*. New York: Twelve / Hachette Book Group, 5.
25. Quoted in: Johnston, George Sim. 2008. "A Melancholy Man of Letters. Review of Samuel Johnson: A Biography by Peter Martin." *Wall Street Journal*, September 18.
26. Fisher, Helen. 2004. *Why We Love: The Nature and Chemistry of Romantic Love*. New York: Henry Holt; Fisher, Helen. 2016/1992. *The Anatomy of Love*. New York: W. W. Norton.
27. Texas Department of Criminal Justice final statements may be found here: http://bit.ly/1oDainR
28. These 7 women represent 1.3% of the 537 executed prisoners in Texas, not very dissimilar to what Cunningham and Vigen (2002) report for national long-term percentages: "2.8% (561/20,000) of the individuals executed in the United States since 1608 have been females, with over half of these executed in the South Census Region. As of December 31, 2000, there had been 45 executions of women in the U.S. since 1900, only 0.56% (45/8,010) of the total executed across this century. During the modern era (post *Furman v. Georgia*, 1972), 137 females have been sentenced to death, representing 23 state jurisdictions. Only five of these women (0.6%) have been among the 683 capital offenders executed in the U.S. since 1973. Of the total of 137 females sentenced to death since 1973, only 53 remained on death row at the close of 2000. Women are under-represented on death row, even in relationship to their rate of arrest for murder. Strieb (2001a) reported that women account for 13% of murder arrests, but only 1.9% of death sentences imposed at trial."
29. Hirschmüller, Sarah, and Boris Egloff. 2016. "Positive Emotional Language in the Final Words Spoken Directly Before Execution." *Frontiers in Psychology*, January, http://bit.ly/1WVnB0h
30. Pennebaker, J. W., C. K. Chung, M. Ireland, A. Gonzales, and R. J. Booth. 2007. *The Development and Psychometric Properties of LIWC2007*. Austin, Texas: LIWC.net.
31. Kashdan, T. B., et al. 2014. "More than Words: Contemplating Death Enhances Positive Emotional Word Use." *Personality and Individual Differences* 71: 171–175.

32. Handelman, L. D., and D. Lester. 2007. "The Content of Suicide Notes from Attempters and Completers." *Crisis* 28: 102–104.

33. Joiner, Thomas. 2006. *Why People Die by Suicide*. Cambridge, MA: Harvard University Press.

34. Carstensen, L. L., D. Isaacowitz, and S. T. Charles. 1999. "Taking Time Seriously. A Theory of Socioemotional Selectivity." *American Psychologist* 54: 165–181.

35. See also: Schuck, Andreas R. T., and Janelle Ward. 2006. "Dealing with the inevitable: strategies of self-presentation and meaning construction in the final statements of inmates on Texas death row." *Discourse and Society* 19(1): 43–62. See also: Kelly, B. D., and S. R. Foley. 2013. "Love, spirituality, and regret: Thematic analysis of last statements from Death Row, Texas (2006–2011)." *Journal of the American Academy of Psychiatry and the Law* 41(4), 540–550. And: Johnson, R., L. C. Kanewske, and M. Barak. 2014. "Death row confinement and the meaning of last words." *Laws* 3(1): 141–152.

36. The average age of the Texas inmates was 39, they waited a mean of 11 years on death row, and their mean educational level was tenth grade. This is in keeping with national longitudinal data on death row inmates. At the end of 2015 there were 2,959 inmates awaiting execution in the United States, 98.18 percent of whom were male. For a good general discussion of demographic data, clinical studies, and research based on institutional records of death row inmates in the United States, see: Cunningham, Mark D., and Mark P. Vigen. 2002. "Death Row Inmate Characteristics, Adjustment, and Confinement: A Critical Review of the Literature." *Behavioral Sciences and the Law* 20: 191–210. They write: "Our analysis shows that death row inmates are overwhelmingly male and disproportionately Southern. Racial representation remains controversial. [They cite these data in the paper: white 45.57%, African American 42.98%, Latino/Latina 9.27%, Native American 1.08%, and Asian 1.08%.] Frequently death row inmates are intellectually limited and academically deficient. Histories of significant neurological insult are common, as are developmental histories of trauma, family disruption, and substance abuse. Rates of psychological disorder among death row inmates are high, with conditions of confinement appearing to precipitate or aggravate these disorders."

37. One study, for example, found that "children from single-parent families are more likely to have behavioral problems because they tend to lack economic security and adequate time with parents" and that "the most reliable indicator of violent crime in a community is the proportion of fatherless families." See: Maginnis, R. L. 1997. "Single-parent Families Cause Juvenile Crime." In *Juvenile Crime: Opposing Viewpoints*, A. E. Sadler, ed., 2012, Greenhaven Press, 62–66.

38. Shermer, Michael. 2016. "Death Wish: What Would Be Your Final Words?" *Scientific American*, June. http://bit.ly/289mKAC

39. http://bit.ly/1srYmdd. Luis Camnitzer's exhibition was also, in part, a commentary on the death penalty, partially "informed by his first-hand experience of dictatorships in Latin America," according to the gallery caption. "This exhibition is particularly timely because it comes on the heels of New Jersey's historic decision to abolish the death penalty, and as the Supreme Court continues to consider the constitutionality of lethal injections." I asked Luis if he'd like to expand on this motivation and he replied: "Capital punishment is premeditated murder. It is the culminating act of an unethical pedagogical project where punishment is confused with education and, in its extreme version, favors elimination over rehabilitation."

40. Pigment prints in 6 parts. 66h × 44w inches (167.64h × 111.76w cm). Courtesy of Alexander Gray Associates, New York © 2016 Luis Camnitzer / Artists Rights Society (ARS), New York.

41. Upton, Maranda A., Tabitha M. Carwile, and Kristina S. Brown. 2016. "In Their Own Words: A Qualitative Exploration of Last Statements of Capital Punishment Inmates in the State of Missouri, 1995–2011." *OMEGA—Journal of Death and Dying*, 1–19. http://ti.me/2cWA575

42. We excluded those who said they were sorry for themselves, or who said that they forgave the prison warden, the state of Texas, and even the victims' families present in the room.

43. Although there were a handful of Muslims who referenced Allah (e.g., reciting the Islamic creed "There is no god but God and Muhammad is his prophet"), there was nothing in the statements to indicate any other religious affiliation.

44. Rosenblatt, A., J. Greenberg, S. Solomon, T. Pyszczynski, and D. Lyon. 1992. "Evidence for Terror Management Theory I: The Effects of Mortality Salience on Reactions to Those Who Violate or Uphold Cultural Values." *Journal of Personality and Social Psychology* 57(4): 681–690.

45. Boehm, Christopher. 2012. *Moral Origins: The Evolution of Virtue, Altruism, and Shame*. New York: Basic Books.

CHAPTER 2

1. I wrote an essay about my mother titled "Shadowlands" that was published in: Shermer, Michael. 2005. *Science Friction*. New York: Henry Holt, 101–109.

2. Nuland, Sherwin B. 1993. *How We Die: Reflections on Life's Final Chapter*. New York: Random House.

3. Slaughter, V., and M. Griffiths. 2007. "Death Understanding and Fear of Death in Young Children." *Clinical Child Psychology and Psychiatry*, 12(4): 525–535.

4. Bering, Jesse M., and David F. Bjorklund. 2004. "The Natural Emergence of Reasoning about the Afterlife as a Developmental Regularity." *Developmental Psychology*, 40(2): 217–233.

5. My goal here is not to provide a module on how to deal with children who have experienced the loss of a loved one—for that I recommend the book by John James and Russell Friedman, *When Children Grieve* (New York: Harper Collins) and the many resources at their Grief Recovery Institute: http://bit.ly /1VsBW7f. See also: Speece, Mark W. 1995. "Children's Concepts of Death." *Michigan Family Review* 01/1: 57–69.

6. James and Friedman, *When Children Grieve*, 227.

7. For a general review of the literature see: Corr, Charles A., and Donna M. Corr, eds. 1996. *Handbook of Childhood Death and Bereavement*. New York: Springer.

8. Slaughter and Griffiths, 2007, op cit.

9. Shtulman, Andrew. 2017. *Scienceblind: Why Our Intuitive Theories About the World Are So Often Wrong*. New York: Basic Books, 146.

10. Reiss, Diana. 2012. *The Dolphin in the Mirror: Exploring Dolphin Minds and Saving Dolphin Lives*. New York: Mariner Books.

11. Alves, F., C. Nicolau, A. Dinis, C. Ribeiro, and L. Freitas. 2014. "Supportive Behavior of Free-Ranging Atlantic Spotted Dolphins (*Stenella frontalis*) Toward Dead Neonates, with Data on Perinatal Mortality." *Acta Ethologica*. doi:10.1007/s10211-014-0210-8.

12. Quoted in: Bates, Mary. 2015. "Do Dolphins Grieve?" *Wired*, January 26. http://bit.ly/1JuVBeg

13. Quoted in: Hooper, Rowan. 2011. "Death in Dolphins: Do They Understand They Are Mortal?" *New Scientist*, August 31. http://bit.ly/1pTwkGe

14. Ibid.

15. Personal correspondence, December 21, 2016. See also: McComb, Karen, Lucy Baker, and Cynthia Moss. 2006. "African Elephants Show High Levels of Interest in the Skulls and Ivory of Their Own Species." *Biology Letters* 2(1): 26–28. ISSN 1744-9561

16. Moss, C. 1988. *Elephant Memories: Thirteen Years in the Life of an Elephant Family*. New York: Fawcett Columbine.

17. See for example: Masson, Jeffrey M., and S. McCarthy. 1995. *When Elephants Weep: The Emotional Lives of Animals*. New York: Delacorte Press; Boehm, C. 1999. *Hierarchy in the Forest: The Evolution of Egalitarian Behavior*. Cambridge, MA: Harvard University Press; Ridley, M. 1997. *The Origins of Virtue: Human Instincts and the Evolution of Cooperation*. New York: Viking.

18. Personal correspondence, April 17, 2016. I am sorry to note that shortly after our correspondence, Russell was diagnosed with cancer; he died in November 2016. He was a friend and colleague whom I shall miss.

19. Solecki, Ralph S. 1975. "The Shanidar IV, a Neanderthal Flower Burial in Northern Iraq." *Science* 190(4217): 880–881.

20. Sommer, Jeffrey D. 1999. "The Shanidar IV 'Flower Burial': A Re-evaluation of Neanderthal Burial Ritual." *Cambridge Archaeological Journal* 9(1): 127–129.

21. Rendu, William et al. 2013. "Evidence Supporting an Intentional Neandertal Burial at La Chapelle-aux-Saints." *PNAS* 111 (1): 81–86. http://bit.ly /2OZxuka

22. Zilhao, J., et al. 2010. "Symbolic Use of Marine Shells and Mineral Pigments by Iberian Neandertals." *PNAS* 107 (3): 1023–1028.

23. Duday, H. 2011. *The Archaeology of the Dead*. Lectures in Archaeothanatology. Oxford: Oxbow Books.

24. Lieberman, Philip. 1991. *Uniquely Human: The Evolution of Speech, Thought, and Selfless Behavior*. Cambridge, MA: Harvard University Press, 163.

25. Riel-Salvatore, Julien, and Claudine Gravel-Miguel. 2013. "Upper Paleolithic Mortuary Practices in Eurasia: A Critical Look at the Burial Record." In *The Oxford Handbook of the Archaeology of Death and Burial*, edited by Sarah Tarlow and Liv Nilsson Stutz, 303–346. Oxford: Oxford University Press.

26. Trinkaus, Erik, Alexandra Buzhilova, Maria Mednikova, and Marvia Dobrovolskaya. 2014. *The People of Sunghir: Burials, Bodies, and Behavior in the Early Upper Paleolithic*. Oxford: Oxford University Press.

27. Pettitt, Paul, Michael Richards, Roberto Maggi, and Vincenzo Formicola. 2003. "The Gravettian Burial Known as the Prince ('Il Principe'): New Evidence for His Age and Diet." *Antiquity* 77: 15–19.

28. See also: Tarlow, Sarah, and Liv Nilsson Stutz, eds. 2013. *The Oxford Handbook of the Archaeology of Death and Burial*. Oxford: Oxford University Press; Pearson, Mike Parker. 2000. *The Archaeology of Death and Burial*. College Station: Texas A&M University Press.

29. Dirks, Paul H.G.M., Lee R. Berger, Eric M. Roberts, et al. 2015. "Geological and Taphonomic Context for the New Hominin Species *Homo naledi* from the Dinaledi Chamber, South Africa." *eLife*. http://bit.ly/1QLswxK

30. http://reut.rs/1ZbdYML

31. http://to.pbs.org/1UHoT08

32. Shermer, Michael. 2016. "Murder in the Cave: Did *Homo naledi* behave more like *Homo homicidensis*?" *Scientific American*, January 1.

33. The paleoanthropologists concluded that "deliberate body disposal" was the likeliest explanation for the find after rejecting four other hypotheses: *Occupation*. There is no debris in the chamber, which is so dark that habitation would have required artificial light, for which there is no evidence, and the cave is nearly inaccessible and appears never to have had easy access. *Water transport*. Caves that have been inundated show sedimentological layers of coarse-grained material, which is lacking in the Dinaledi Chamber where the *H. naledi* specimens were uncovered. *Predators*. There are no signs of predation on the skeletal remains and no fossils from predators. *Death trap*. The sedimentary remains indicate that the fossils were deposited over a span of

time, so that rules out a single calamitous event, and the near unreachability of the chamber makes attritional individual entry and death unlikely.

CHAPTER 3

1. http://bit.ly/1DMBgwJ
2. Quoted in: Olan, Levi A. 1971. *Judaism and Immortality*. New York: Union of American Hebrew Congregations, 66.
3. Eisenman, Robert H., and Michael Wise. 1992. *The Dead Sea Scrolls Uncovered*. Rockport: Element, 20.
4. Freedman, David Noel, editor in chief. 1992. *The Anchor Bible Dictionary*. New York: Doubleday. vol. 3, 90–93.
5. Ibid., 93.
6. John Paul II. 1999. "General Audience." July 21. http://bit.ly/29fK0G5
7. For an excellent succinct discussion of Islam and the afterlife see: Segal, Alan. 2004. *Life After Death: A History of the Afterlife in Western Religion*. New York: Random House, 639–695.
8. Hirsi Ali, Ayaan. 2015. *Heretic: Why Islam Needs a Reformation Now*. New York: HarperCollins, 13.
9. Murata, Sachiko, and William Chittick. 1998. *The Vision of Islam*. London: I.B. Tauris, 167–168.
10. Ibid., 169.
11. Watt, W. M., A. J. Wensinck, R. B. Winder, and D. A. King. 2009. "Makka." *Encyclopedia of Islam*, 2nd ed. Leiden: E. J. Brill. vol. 6, 144.
12. Wright, J. Edward. 2000. *The Early History of Heaven*. New York: Oxford University Press.
13. McGrath, Alister E. 2003. *A Brief History of Heaven*. Oxford: Blackwell Publishing.
14. Russell, Jeffrey Burton. 1997. *A History of Heaven: The Singing Silence*. Princeton, NJ: Princeton University Press.
15. Segal, Alan. 2004. Op. cit.
16. McDannell, Colleen, and Bernhard Lang. 1988. *Heaven: A History*. New Haven: Yale University Press. See also the highly readable engaging account of the concept in: Miller, Lisa. 2010. *Heaven: Our Enduring Fascination with the Afterlife*. New York: HarperCollins.
17. For good measure see Alice K. Turner's *The History of Hell*, Elaine Pagels's *The Origin of Satan*, and Alan Bernstein's *The Formation of Hell*, not to mention Jacques Le Goff's *The Birth of Purgatory*, for a sampling of the history of the other place people think only other people are going to after death. Turner, Alice K. 1993. *The History of Hell*. New York: Harcourt Brace; Pagels, Elaine. 1995. *The Origin of Satan*. New York: Random House; Bernstein, Alan E. 1993. *The Formation of Hell: Death and Retribution in*

the Ancient and Early Christian Worlds. Ithaca: Cornell University Press; Le Goff, Jacques. 1981. *The Birth of Purgatory.* Chicago: University of Chicago Press.

18. Segal, 2004, 698.

19. McDannell and Lang, xxii.

20. Gallup, George, Jr., and James Castelli. 1989. *The People's Religion: American Faith in the 90's.* New York: Macmillan.

21. 2008. "Many Americans Say Other Faiths Can Lead to Eternal Life." Pew Forum. December 18. http://pewrsr.ch/29dfI6M

22. Moses Maimonides. 1170–1180. "The Blissful State of the Soul." *Mishneh Torah: The Book of Knowledge.* 1965. Jerusalem: Boys Town Jerusalem Publishers, 367.

23. Sweeney, Julia. 2006. *Letting Go of God.* http://bit.ly/1HgJnZ9

24. Quoted in: Cave, Stephen. 2012. *Immortality: The Quest to Live Forever and How it Drives Civilization.* New York: Crown, 60–61.

25. Lamb, Brian. 2002. Interview with Christopher Hitchens and Andrew Sullivan. Genius.com. http://genius.com/1914401

26. Patel, Hasan Salim. 2011. "Christopher Hitchens: A Life in Quotes." *Aljazeera,* December 16. http://bit.ly/1nqGTPS

27. Hitchens, Christopher. 2011. "Is There an Afterlife?" UCLA Debate, February 15. In Hitchens's case, as the end neared, he reflected on "a long argument I am currently having with the specter of death. Nobody ever wins this argument, though there are some solid points to be made while the discussion goes on. I have found, as the enemy becomes more familiar, that all the special pleading for salvation, redemption and supernatural deliverance appears even more hollow and artificial to me than it did before." This quote is from his acceptance speech for the 2011 Richard Dawkins Award. Texas Freethought Convention. http://bit.ly/1VS6UVg

28. The quote appears to have originated with someone less famous than Woody Allen, but as with most quotes it gravitated up to the most famous person who said it, the latest being Stephen Hawking, who at least attributed it to Allen, but it is only a matter of time before the attribution is dropped.

29. *The Interpreter's Bible: The Holy Scriptures in the King James and Revised Standard Versions with General Articles and Introduction, Exegesis, Exposition for Each Book of the Bible.* George Arthur Buttrick, ed. 1957. Nashville: Abingdon Press. vol. 5, 755.

CHAPTER 4

1. Jalal al-Din Rumi, Maulana. 2011. *Selected Poems of Rumi.* Translated from Persian by Reynold A. Nicholson. New York: Dover, 43.

2. Bloom, Paul. 2004. *Descartes' Baby: How the Science of Child Development Explains What Makes Us Human.* New York: Basic Books.

3. Bloom, Paul, and Dave Pizarro. 1999. "Homer's Soul." *The Psychology of the Simpsons*. Brown, Alan S., and Chris Logan, eds. Smart Pop, 65–73.
4. Bloom, Paul. 2004. "Natural-Born Dualists." *Edge.org*. http://bit.ly/29Dm54Y
5. Robinson, Howard, and Edward N. Zalta. 2011. "Dualism." *The Stanford Encyclopedia of Philosophy*. http://stanford.io/2ajlwKZ
6. Chopra, Deepak, and Menas Kafatos. 2016. "Reality Gets an Unlikely Savior: Infinity." *San Francisco Gate*, July 3. http://bit.ly/29jezgN
7. Chopra, Deepak, and Menas Kafatos. 2017. *You Are the Universe*. New York: Harmony Books, 249.
8. van Lommel, Pim. 2010. *Consciousness Beyond Life: The Science of the Near-Death Experience*. New York: HarperCollins.
9. Kelly, Edward F., and Emily Williams Kelly. 2009. *Irreducible Mind: Toward a Psychology for the 21st Century*. Lanham, MD: Rowman and Littlefield, 28.
10. Martin, Michael, and Keith Augustine, eds. 2015. *The Myth of an Afterlife: The Case against Life after Death*. Lanham, MD: Rowman and Littlefield.
11. Larson, E. J., and L. Witham. 1998. "Leading Scientists Still Reject God." *Nature* 394: 313.
12. The German philosopher Ludwig Wittgenstein, considered by many to be the most influential philosopher of the twentieth century, persuasively argued that language, thought, and reality cannot be separated. In his groundbreaking 1921 work, *Tractatus Logico-Philosophicus*, Wittgenstein laid out his propositions that "the world is everything that is the case," that facts about the world are represented in thoughts, then propositions, and finally as language. The words we use to describe the facts about world shape, or even determine how we think about the world itself. "The picture is a model of reality," he wrote, and that picture is described with words. Thus, Wittgenstein famously concluded, "whereof one cannot speak, thereof one must be silent."
13. Pennycook, Gordon, James Allan Cheyne, et al. 2015. "On the Reception and Detection of Pseudo-profound Bullshit." *Judgment and Decision Making* 10(6): 549–563.
14. http://bit.ly/1PQqk6s
15. Personal correspondence, February 22, 2016.
16. Epel, E. S., E. Puterman, J. Lin, E. H. Blackburn, P. Y. Lum, N. D. Beckmann, J. Zhu, E. Lee, A. Gilbert, R. A. Rissman, R. E. Tanzi, and E. E. Schadt. 2016. "Meditation and vacation effects have an impact on disease-associated molecular phenotypes." *Translational Psychiatry* 6, e880; doi:10.1038/tp.2016.164. www.nature.com/tp
17. Personal correspondence, September 8, 2016.
18. This research was given an additional boost in a second study published in 2016 on the effects of a vegetarian diet, meditation, yoga, and massage on a set of metabolic biomarkers associated with inflammation, cardiovascular disease risk, and cholesterol regulation in 119 men and women participants between 30 and 80 years of age enrolled at the Chopra Center. Blood plasma

analyses before and after the six-day program revealed that the half who went through the program showed measurable decreases "in 12 specific cell-membrane chemicals" related to "inflammation and cholesterol metabolism," both of which are "highly predictive of cardiovascular disease." Again, it is difficult to tell what is driving the salutary effects: (a) diet, (b) meditation, (c) yoga, (d) massage, or (e) all of the above? I suspect the answer is (e). So while one's state of mind unquestionably has some measurable effect on one's state of health, I remain skeptical that "consciousness" is a driving factor, however that would be defined and measured. Peterson, Christine Tara; Lucas, Joseph; St. John Williams, Lisa; Thompson, J. Will; Moseley, M. Arthur; Patel, Sheila; Peterson, Scott; Porter, Valencia; Schadt, Eric E.; Mills, Paul J; Tanzi, Rudolph E.; Doraiswamy, P. Murali; and Chopra, Deepak. 2016. "Identification of Altered Metabolomic Profiles Following a Panchakarma-based Ayurvedic Intervention in Healthy Subjects." *Nature/Scientific Reports* 6.32609. doi:10.1038/srep32609.

19. Chopra, Deepak. 2015. *The Human Universe*. http://bit.ly/1VSFaRA
20. Chopra, Deepak. 2006. *Life After Death: The Burden of Proof*. New York: Harmony, 26–27.

CHAPTER 5

1. Kundera, Milan. 1990. *Immortality*. New York: HarperCollins, 215.
2. FAQ International Association for Near Death Studies. http://bit.ly/1Z0LlkR
3. Atwater, P.M.H. 1992. "Is There a Hell? Surprising Observations about the Near-Death Experience." *Journal of Near-Death Studies* 10(3).
4. Schoonmaker, Fred. 1979. "Denver Cardiologist Discloses Findings After 18 Years of Near-Death Research." *Anabiosis* 1:1–2.
5. Gallup, George, Jr. 1982. *Adventures in Immortality*. New York: McGraw-Hill.
6. Van Lommel, P., R. V. Wees, V. Meyers, and I. Elfferich. 2001. "Near-Death Experience in Survivors of Cardiac Arrest: A Prospective Study in the Netherlands." *Lancet* 358(9298): 2039.
7. Van Lommel, P. 2010. *Consciousness Beyond Life: The Science of Near-Death Experience*. New York: HarperCollins.
8. Reported in: Holden, Janice, Bruce Greyson, and Debbie James. 2009. *The Handbook of Near-Death Experiences: Thirty Years of Investigation*. Santa Barbara: Praeger.
9. Hume, David. [1758] 1952. *An Enquiry Concerning Human Understanding*. Great Books of the Western World. Chicago: University of Chicago Press, 491.
10. Woods, Mark. 2015. "'The Boy Who Came Back from Heaven': Alex Malarkey says best-selling book is false." *Christianity Today*, January 15. http://bit.ly/1u6Dkke

11. Lichfield, Gideon. 2015. "The Science of Near-Death Experiences." *Atlantic*, April. http://theatln.tc/21HeWmB

12. Van Lommel et al., 2001.

13. Crislip, Mark. 2008. "Near Death Experiences and the Medical Literature." *Skeptic* 14(2): 14–15. As for "brain death," most people have no idea what this actually means in a medical sense; Dr. Crislip clarifies it thus:

    > The patient has to have no clinical evidence of brain function by physical examination, including no response to pain and a variety of nerve reflexes that do not work: cranial nerve pupillary response (fixed pupils), oculocephalic reflex (steady gaze), corneal reflex (lack of reflexive blinking to stimulation), and no spontaneous respirations. They have to be off all drugs that mimic brain death for several days and they cannot have metabolic conditions that mimic death. It is important to distinguish between brain death and states that mimic brain death and most of the patients received either a benzo [Valium-like drug] and/or a narcotic. A flat line EEG, two at least 24 hours apart, is another criterion. In other words, being declared brain dead is a time-consuming and detailed procedure, as it should be.

14. Alexander, Eben. 2012. *Proof of Heaven: A Neurosurgeon's Journey Into the Afterlife*. New York: Simon and Schuster; Alexander, Eben. 2012. "Proof of Heaven: A Doctor's Experience with the Afterlife." *Newsweek*, October 8. http://bit.ly/1pCWqX5

15. Harris, Sam. 2014. *Waking Up: A Guide to Spirituality Without Religion*. New York: Simon and Schuster, 3–5.

16. Quoted in: Shermer, Michael. 2011. *The Believing Brain*. New York: Henry Holt, 13.

17. Quoted in: Pollan, Michael. 2015. "The Trip Treatment." *New Yorker*, February 9, 38.

18. Sacks, Oliver. 2015. *On the Move: A Life*. New York: Random House, 142.

19. Quoted in: Gottlieb, Robert. 2014. "To Heaven and Back!" *New York Review of Books*, October 23. http://bit.ly/2hxkhen

20. Sacks, Oliver. 2012. *Hallucinations*. New York: Alfred A. Knopf.

21. Sacks, Oliver. 2012. "Seeing God in the Third Millennium." *Atlantic*, December 12. http://theatln.tc/1bP4lK4

22. Blackmore, Susan. 1993. *Dying to Live: Near-Death Experiences*. Buffalo, NY: Prometheus Books.

23. In Sacks, 2012. *Hallucinations*.

24. Blanke, O., S. Ortigue, T. Landis, and M. Seeck. 2002. "Neuropsychology: Stimulating Illusory Own-body Perceptions." *Nature* 419, September 19: 269–270.

25. Whinnery, J. E., and A. M. Whinnery. 1990. "Acceleration-Induced Loss of Consciousness: A Review of 500 Episodes." *Archives of Neurology* 47: 764–776.

26. Whinnery, J. E. 1979. "Technique for Simulating G-Induced Tunnel Vision." *Aviation and Space Environmental Medicine* 50: 1076.
27. Markum, Cory. 2014. "Heaven Only Knows: Near-Death Experiences and the Problem of Account Incongruence." *Skeptic* 20(2): 12–15.
28. Bhagavad Gita. Translated from the Sanskrit by Juan Mascaró. 1962. New York: Penguin, 10.
29. Ibid., 11.
30. Goldberg, Bruce. 1982. *Past Lives, Future Lives: A Hypnotherapist Shares His Most Astounding Case Histories*. New York: Ballantine Books, 181.
31. Goldberg, Bruce. "Time Travelers I Have Met." http://bit.ly/1WSuNOa
32. Loftus, Elizabeth, and J. C. Palmer. 1974. "Reconstruction of Automobile Destruction: An Example of the Interaction between Language and Memory." *Journal of Verbal Learning and Verbal Behavior* 13: 585–589.
33. Loftus, Elizabeth, C. Manning, and S. J. Sherman. 1996. "Imagination Inflation: Imagining a Childhood Event Inflates Confidence That It Occurred." *Psychonomic Bulletin and Review* 3: 208–214.
34. Spanos, Nicholas. 1996. *Multiple Identities and False Memories: A Socio-cognitive Perspective*. Washington, DC: American Psychological Association, 135–140.
35. Stevenson, Ian. 1997. *Reincarnation and Biology*. 2 vols. Westport, CT: Praeger.
36. Angel, Leonard. 2003. "Reincarnation All Over Again: Backwards Reasoning in Ian Stevenson's *Reincarnation and Biology*." *Skeptic* 9(3).
37. Edwards, Paul. 1996. *Reincarnation: A Critical Examination*. Buffalo: Prometheus Books, 28, 255.
38. The show we all appeared on was *Larry King Live*. The book about their case is: Leininger, Bruce, and Andrea Leininger. With Ken Gross. 2009. *Soul Survivor: The Reincarnation of a World War II Fighter Pilot*. New York: Grand Central Publishing.
39. Transcript of *Larry King Live* episode on CNN, December 22, 2009: http://cnn.it/1UdkKNp
40. Quoted in: "The Past Life Memories of James Leininger." *Facts Are Facts*, http://bit.ly/1sG65nC
41. Quoted in: "Parents Think Boy Is Reincarnated Pilot." 2015. *ABC News*, June 30. http://abcn.ws/1qVGTs3
42. Quoted in: "The Past Life Memories of James Leininger." *Facts Are Facts*, http://bit.ly/1sG65nC
43. Quoted in: "Parents Think Boy Is Reincarnated Pilot." 2015. *ABC News*, June 30. http://abcn.ws/1qVGTs3
44. Quoted in: "The Past Life Memories of James Leininger." *Facts Are Facts*, http://bit.ly/1sG65nC

45. http://bit.ly/1TGXyrs
46. Bowman explains this memory loss by claiming that there is just a short window of time in childhood, from age two or three to six or seven, when past life memories are revealed, after which they are gone.

CHAPTER 6

1. Sagan, Carl. 1996. *The Demon-Haunted World: Science as a Candle in the Dark*. New York: Random House, 203.
2. Discussed in detail in: Shermer, Michael. 1997. *Why People Believe Weird Things*. New York: Henry Holt; Shermer, Michael. 2011. *The Believing Brain*. New York: Henry Holt.
3. Quoted in: Woerlee, G. M. 2005. *Mortal Minds: The Biology of Near Death Experiences*. Buffalo, NY: Prometheus Books, 95–96.
4. Kripal, Jeffrey J. 2014. "Visions of the Impossible." *The Chronicle of Higher Education*, March 31. http://bit.ly/2jVZHcq
5. Reed, Graham. 1988. *The Psychology of Anomalous Experience*. Buffalo, NY: Prometheus Books.
6. Zusne, Leonard, and Warren H. Jones. 1989. *Anomalistic Psychology: A Study of Magical Thinking*. New York: Lawrence Erlbaum Associates.
7. Cardena, Etzel, Steven Jay Lynn, and Stanley Krippner, eds. 2000. *Varieties of Anomalous Experience: Examining the Scientific Evidence*. Washington, DC: American Psychological Association, 4.
8. Van Praagh, James. 2011. *Growing Up in Heaven*. New York: HarperOne, 10.
9. Ibid., 94, 18.
10. Filmed in a television studio in Seattle in 2002. Watch the episode here: http://bit.ly/2iUmST2
11. www.thecoldreadingconnection.com
12. Schwartz, Gary. 2002. *The Afterlife Experiments: Breakthrough Scientific Evidence of Life After Death*. New York: Atria Books.
13. Berard, Marc. 2003. "I See Dead People. Review of *The Afterlife Experiments*." *Skeptic* 9(3).
14. Hyman, Ray. 2003. "How Not to Test Mediums: Critiquing the Afterlife Experiments." *Skeptical Inquirer* 27(1), January. http://bit.ly/2koCmPf
15. See: Tavris, Carol, and Elliot Aronson. 2007. *Mistakes Were Made (but Not by Me): Why We Justify Beliefs, Bad Decision, and Hurtful Acts*. New York: Houghton Mifflin.
16. Newberg, Andrew, and Eugene D'Aquili. 2001. *Why God Won't Go Away: Brain Science and the Biology of Belief*. New York: Random House; Newberg, Andrew, and Mark Robert Waldman. 2006. *Born to Believe: God, Science and the Origin of Ordinary and Extraordinary Beliefs*. New York:

Free Press; Newberg, Andrew, and Mark Robert Waldman. 2009. *How God Changes Your Brain*. New York: Ballantine Books.

17. Peres, Julio Fernando, Alexander Moreira-Almeida, Leonardo Caixeta, Frederico Leao, and Andrew Newberg. 2012. "Neuroimaging during Trance State: A Contribution to the Study of Dissociation." *PLoS*, November 16. http://bit.ly/2jZZkgS

18. See also: Peres, Julio F. P., and Andrew Newberg. 2013. "Neuroimaging and Mediumship: A Promising Research Line." *Archives of Clinical Psychiatry* 40(6). http://bit.ly/2kKggYk

19. Houdini, Harry. 1924/2011. *A Magician Among the Spirits*. New York: Cambridge University Press.

20. Shermer, Michael. 2014. "Infrequencies." *Scientific American*, October. http://bit.ly/1rGc4qd

21. Sagan, 1996, 104.

22. Coyne, Jerry. 2014. "Science Is Being Bashed by Academics Who Should Know Better." *New Republic*, April 3. http://bit.ly/2jVY3CT. See Kripal's response to Coyne's critique here: Kripal, Jeffrey. 2014. "Embracing the Unexplained, Part 2." *The Chronicle of Higher Education*, April 8. http://bit.ly/2kkTeGt

23. Quoted in: Gleick, James. 1993. *Genius: The Life and Science of Richard Feynman*. New York: Vintage, 93.

24. Thorne, Kip. 2014. *The Science of Interstellar*. Los Angeles: Warner Brothers, 264.

CHAPTER 7

1. "Second Chances." 1993. *Star Trek, The Next Generation*. Episode 150. Aired May 24. Summary here: http://bit.ly/1SdFwvv. Script here: http://bit.ly/1RvhleL

2. Chisholm, Roderick M. 2004. *Person and Object: A Metaphysical Study*. London: Routledge, 89.

3. Wenner, Melinda. 2007. "Humans Carry More Bacterial Cells than Human Ones." *Scientific American*, November 30. http://bit.ly/1uhlM0s

4. Margulis, Lynn. 1998. *Symbiotic Planet: A New Look at Evolution*. New York: Basic Books; Margulis, Lynn. 2011. "Symbiogenesis. A new principle of evolution rediscovery of Boris Mikhaylovich Kozo-Polyansky (1890–1957)." *Paleontological Journal* 44(12): 1525–1539.

5. Gordon, L., et al. 2012. "Neonatal DNA methylation profile in human twins is specified by a complex interplay between intrauterine environmental/genetic factors subject to tissue-specific influence." *Genome Research* 22: 1395–1406.

6. Lodato, Michael A., et al. 2015. "Somatic Mutation in Single Human Neurons Tracks Developmental and Transcriptional History." *Science* 350(6256): 94–98. http://bit.ly/1U0XtCS

7. McConnell, M. J., et al. 2013. "Mosaic Copy Number Variation in Human Neurons." *Science* 342(6158): 632–637. http://1.usa.gov/1NwnO5d

8. Quoted in: Yong, Ed. 2015. "The Surprising Genealogy of Your Brain." *Atlantic*, October 1, http://theatln.tc/1YUtuOa

9. Flanagan, Owen. 2002. *The Problem of the Soul: Two Visions of Mind and How to Reconcile Them.* New York: Basic Books.

10. Flanagan, 2002, 165–166.

11. Ibid., 164.

12. Kurzban, Robert. 2012. *Why Everyone (Else) Is a Hypocrite.* Princeton, NJ: Princeton University Press.

13. Research in cognitive psychology also supports this proposition, elegantly summarized by the cognitive psychologist Bruce Hood in *The Self Illusion*, employing an analogy with a science fiction film: "We process the outside world through our nervous system in order to create a model of reality in our brains. And, just like the matrix in the science fiction movie, not everything is what it seems. We all know the power of visual illusions to trick the mind into perceiving things incorrectly, but the most powerful illusion is the sense that we exist inside our heads as an integrated, coherent individual or self." Hood, Bruce. 2012. *The Self Illusion: How the Social Brain Creates Identity.* Oxford: Oxford University Press, 3.

14. Kestenbaum, David. 2007. "Atomic Tune-Up: How the Body Rejuvenates Itself." NPR, July 14, http://n.pr/1qbBP3a

15. Wade, Nicholas. 2005. "Your Body Is Younger Than You Think." *New York Times*, August 2. http://nyti.ms/1pLXzC4

16. Jabr, Ferris. 2012. "Know Your Neurons: What Is the Ratio of Glia to Neurons in the Brain?" *Scientific American*, June 13. http://bit.ly/1W3nJeF

17. Shermer, Michael. 1995. "Exorcising Laplace's Demon: Chaos and Anti-chaos, History and Metahistory." *History and Theory* 34(1): 59–83.

18. Tipler, Frank J. 1994. *The Physics of Immortality: Modern Cosmology, God, and the Resurrection of the Dead.* New York: Doubleday.

19. The difference in the spelling—googolplex and googleplex—is because, says Google cofounder Larry Page, they didn't yet have spell check when they named the company.

CHAPTER 8

1. Hochman, David. 2016. "Reinvent Yourself: The Playboy Interview with Ray Kurzweil." *Playboy*, April 19. http://bit.ly/1U6WcKL

2. Personal correspondence, July 30, 2014.

3. Personal correspondence, September 26, 2013.

4. Ibid. To read more go to: www.merkle.com. Merkle's video introduction to molecular nanotechnology is here: http://bit.ly/1MTXJCl

5. Personal correspondence, September 24, 2013.

6. Aaronson, Xavier. Motherboard. 2016. *Frozen Faith: Cryonics and the Quest to Cheat Death*. Vice. Thobey Campion, executive producer. http://bit.ly/1TB47M7

7. http://bit.ly/2nVE1iI

8. More, Max, and Natasha Vita-More, eds. 2013. *The Transhumanist Reader: Classical and Contemporary Essays on the Science, Technology, and Philosophy of the Human Future*. New York: Wiley-Blackwell, 4.

9. Sandberg, Anders. 1999. "The Physics of Information Processing Super-objects: Daily Life Among the Jupiter Brains." *Journal of Evolution and Technology* 5(1).

10. Walker, Mark. 2011. "Personal Identity and Uploading." *Journal of Evolution and Technology* 22(1). http://bit.ly/1SAbXHB

11. http://bit.ly/1ULQv69

12. Bostrom, Nick. 2003. "Are You Living in a Computer Simulation?" *Philosophical Quarterly* 43(211): 243–255. http://bit.ly/19EKA6E

13. Tipler, Frank J. 1994. *The Physics of Immortality: Modern Cosmology, God, and the Resurrection of the Dead*. New York: Doubleday; Tipler, Frank J. 2007. *The Physics of Christianity*. New York: Doubleday.

14. Shermer, Michael. 1997. *Why People Believe Weird Things*. New York: W. H. Freeman.

15. Personal correspondence, September 11, 1995.

16. Barrow, John, and Frank J. Tipler. 1986. *The Anthropic Cosmological Principle*. Oxford: Oxford University Press, 677.

17. Shermer, 1997, 255–272.

18. Krauss, Lawrence. 2007. "More Dangerous than Nonsense." *New Scientist*, May 12. http://bit.ly/1M7WsHw

19. Ptolemy, Barry. 2009. *Transcendent Man,* op. cit.

20. The Singularity Is Near. Questions and Answers. http://bit.ly/1EV4jk0

21. Kurzweil, Ray, and Terry Grossman. 2009. *Transcend: Nine Steps to Living Well Forever*. Rodale Press, xxv.

22. Kurzweil, Ray, and Terry Grossman. 2004. *Fantastic Voyage: Live Long Enough to Live Forever*. Rodale Press, 3.

23. Hochman, 2016, op cit.

24. Anderson, Kyle. 2015. "Google's Larry Page and Sergey Brin Plan to Cure Aging with Biotech Venture." *Money Morning*. http://bit.ly/1djWZ7X

25. For evidence that memories are stored in static synaptic connections see: http://bit.ly/1Wgh1U8

26. 21st Century Medicine website: http://www.21cm.com/

27. Kidney transplant paper: http://1.usa.gov/1QqPwjN

28. The reality of what they were about to do hit me hard, coming as it did on the heels of writing my chapter on animal rights for *The Moral Arc*. I swallowed my moral qualms and reminded myself that biomedical science often depends on animal models, at least for a time, and that such animals would

never have been born were it not for their use in research. It was heartening to see that these scientists were extraordinarily gentle with and sensitive to the well-being of their furry charge.

29. This entire process is called Aldehyde-Stabilized Cryopreservation and is described here: http://bit.ly/1XLsNER

30. Personal correspondence, November 23, 2015.

31. http://bit.ly/1XLsNER

32. Ken Hayworth added: "We have been so fortunate to have two top-notch research labs competing in our prize competition. Over the last year both teams have submitted many, many brain samples for our X-ray and electron microscopic evaluation. And both teams have published peer-reviewed scientific papers that claim successful preservation of intact mammalian brains at the connectome level. Here are links to those two papers:

Mikula publication: http://bit.ly/1pmFABH

21CM publication: http://bit.ly/245ySQ9

I literally had no idea which team would win the prize until quite recently. In recent months, both teams submitted intact brains for official evaluation in the prize contest. I spent considerable time imaging both of these entries, as well as an additional brain entry from Mikula that was designed to address damage seen in a previous entry.

Here are links to pages on our website that display these electron micrographs:

Mikula mouse brain evaluation images: http://bit.ly/1Nn4I6M

21CM rabbit brain evaluation images: http://bit.ly/1U7LnrH

Both entries were of high quality, but the Mikula entry showed some damage to axon tracts in the central regions of the brain, presumably due to chemical penetration issues that were evident as lighter staining in those central regions. It also showed a few small cracks in some peripheral cortical regions. These problems are likely straightforward to address, but they do show clear damage to this brain's connectome, disqualifying the entry according to our rules. In contrast, the 21CM rabbit brain's connectome looks to be essentially damage free, according to our extensive 2D electron microscopy (EM) survey images and three 3DEM volume datasets. Minor fixation artifacts were seen in both entries but were deemed to not disrupt the traceability of the connectome." Shortly after the Small Mammal Brain Preservation Prize was awarded to Robert McIntyre and Greg Fahy, McIntyre left 21st Century Medicine to start his own neuroscience company called Nectome to further develop ACS on his own: www .nectome.com

33. Personal correspondence, October 28, 2015. All quotes from Hayworth in this chapter come from these emails.

34. Sullivan, B. J., L. N. Sekhar, D. H. Duong, G. Mergner, and D. Alyano. 1999. "Profound Hypothermia and Circulatory Arrest with Skull Base Approaches for Treatment of Complex Posterior Circulation Aneurysms." *Acta Neurochir* 141: 1012.

35. Lomber, S.G., B. R. Payne, and J. A. Horel. 1999. "The Cryoloop: An Adaptable Reversible Cooling Deactivation Method for Behavioral or Electrophysiological Assessment of Neural Function." *Journal of Neuroscience Methods* 86: 179–194.

36. Bailey, C. H., E. R. Kandel, and K. M. Harris. 2015. "Structural Components of Synaptic Plasticity and Memory Consolidation." *Cold Spring Harbor Perspectives in Biology*, 1–19.

37. Tonegawa, S., M. Pignatelli, D. S. Roy, and T. J. Ryan. 2015. "Memory Engram Storage and Retrieval." *Current Opinion in Neurobiology* 35: 101–109.

38. The principle has come to be known as the Hebbian theory, after the man who first proposed it: Hebb, Donald. 1949. *The Organization of Behavior*. New York: John Wiley and Sons. The phrase itself comes from Löwel, Siegrid, and W. Singer. 1992. "Selection of Intrinsic Horizontal Connections in the Visual Cortex by Correlated Neuronal Activity." *Science* 255, January 10: 209–212.

39. Kandel, E. R., et al. 2012. *Principles of Neural Science*, 5th ed. New York: McGraw-Hill.

40. Crick, Francis. 1994. *The Astonishing Hypothesis: The Scientific Search for the Soul*. New York: Touchstone, 3.

41. Parfit, Derek. 1984. *Reason and Persons*. Oxford: Oxford University Press, 254–255.

42. Philosophers love thought experiments, so much that the philosopher Dan Dennett includes them in his "intuition pumps" tool kit because they help pump out ideas to challenge our intuitions about the world, which cognitive psychologists have demonstrated are not always reliable.

43. Merkle, Ralph. 2009. "The State of the Art of Cryopreservation." Lecture at the 2009 Longevity Summit. http://bit.ly/1NuGBDh

44. Eddington, Arthur Stanley. 1928. *The Nature of the Physical World*. New York: Macmillan, 74.

45. Quoted in: Webster, Lisa. 2015. "The Promise of Immortality in a Tech-Enhanced Heaven." *Religion Dispatches*, April 24. http://bit.ly/27SKSaE

CHAPTER 9

1. http://bit.ly/1PN6jxt

2. White House Office of the Press Secretary. 2016. "Remarks by President Obama in Address to the People of Europe. Hannover, Germany, April 25." http://bit.ly/1QwUKKA

3. President Barack Obama's final speech before the United Nations. September 21, 2016. http://ti.me/2cWA575

4. See, for example, the economist Max Roser's ourworldindata.org and the data at humanprogress.org aggregated from the World Bank, the UN, OECD, and Eurostat. See also: Norberg, Johan. 2016. *Progress: Ten Reasons to Look Forward to the Future*. London: OneWorld Publications; Diamandis, Peter, and Steven Kotler. 2012. *Abundance: The Future Is Better than You Think*. New York: Free Press; Ridley, Matt. 2011. *The Rational Optimist: How Prosperity Evolves*. New York: HarperCollins; Pinker, Steven. 2011. *The Better Angels of Our Nature*. New York: Penguin; Clark, Gregory. 2007. *A Farewell to Alms: A Brief Economic History*. Princeton, NJ: Princeton University Press; Beinhocker, Eric. 2006. *The Origin of Wealth: Evolution, Complexity, and the Radical Remaking of Economics*. Cambridge, MA: Harvard Business School Press; and my own 2015 book *The Moral Arc: How Science and Reason Lead Humanity Toward Truth, Justice, and Freedom*. New York: Henry Holt.

5. http://bit.ly/1eRbn2E

6. DeLong, J. Bradford. 2000. "Cornucopia: The Pace of Economic Growth in the Twentieth Century." Working Paper 7602. National Bureau of Economic Research. http://bit.ly/2fQxcXn. See also: DeLong, J. Bradford. 1998. "Estimating World GDP, One Million B.C.–Present." http://bit.ly/2fCWomN

7. For a comprehensive economic history of humanity and why wealth has evolved as it has, see: Beinhocker, Eric. 2006. *The Origin of Wealth: Evolution, Complexity, and the Radical Remaking of Economics*. Cambridge, MA: Harvard Business School Press.

8. Clark, Gregory. 2007. *A Farewell to Alms: A Brief Economic History*. Princeton, NJ: Princeton University Press, 2–3.

9. See Max Roser's ourworldindata.org, the data at humanprogress.org aggregated from the World Bank, the UN, OECD, and Eurostat, and Data in Gapminder World that tracks over five hundred areas of change, most in the positive direction: gapminder.org/data/

10. Data source: World Bank and: Bourguignon, Francois, and Christian Morrisson. 2002. "Inequality among World Citizens: 1820–1992." *American Economic Review* 92(4): 727–744.

11. Reported in: Etchells, Pete. 2015. "Declinism: Is the World Actually Getting Worse?" *Guardian*, January 16. http://bit.ly/2cWK1D7

12. Chambers, John R., Lawton K. Swan, and Martin Heesacker. 2013. "Better Off Than We Know: Distorted Perceptions of Incomes and Income Inequality in America." *Psychological Science*, 1–6.

13. Shermer, Michael. 2008. *The Mind of the Market: How Biology and Psychology Shape Our Economic Lives*. New York: Henry Holt.

14. Kirkpatrick, Curry. 1975. "Cool Warmup for Jimbo." *Sports Illustrated*, April 28.

15. http://imdb.to/2ehcqDH

16. Opening scene of *The Armstrong Lie* by Alex Gibney, 2013.

17. Gilovich, Thomas, and Gary Belsky. 2000. *Why Smart People Make Big Money Mistakes and How to Correct Them. Lessons from the New Science of Behavioral Economics.* New York: Fireside.

18. Thaler, Richard. 1980. "Toward a Positive Theory of Consumer Choice." *Journal of Economic Behavior and Organization,* reprinted in Breit and Hochman, eds. *Readings in Microeconomics,* 3rd ed.

19. Thaler experiment: Thaler, Richard, Daniel Kahneman, and Jack Knetsch. 1990. "Experimental Tests of the Endowment Effect and the Coase Theorem." *Journal of Political Economy,* December. http://bit.ly/2fQwJEN

20. Chen, Keith, Venkat Lakshminarayanan, and Laurie Santos. 2006. "How Basic Are Behavioral Biases? Evidence from Capuchin-Monkey Trading Behavior." *Journal of Political Economy,* June.

21. Baumeister, Roy F., Ellen Bratslavsky, Catrin Finkenauer, and Kathleen D. Vohs. 2001. "Bad Is Stronger Than Good." *Review of General Psychology* 5(4): 323–370.

22. Gilbert, A. N., A. J. Fridlund, and J. Sabini. 1987. "Hedonic and Social Determinants of Facial Displays to Odors." *Chemical Senses* 12: 355–363.

23. Rothbart, M., and B. Park. 1986. "On the confirmability and disconfirmability of trait concepts." *Journal of Personality and Social Psychology* 50: 131–142.

24. Bless, H., D. L. Hamilton, and D. M. Mackie. 1992. "Mood effects on the organization of person information." *European Journal of Social Psychology* 22: 497–509; Skowronski, J. J., and D. E. Carlston. 1989. "Negativity and extremity biases in impression formation: A review of explanation." *Psychological Review* 105: 131–142; Dreben, E. K., S. T. Fiske, and R. Hastie. 1979. "The independence of evaluative and item information: Impression and recall order effects in behavior based impression formation." *Journal of Personality and Social Psychology* 37: 1758–1768.

25. Ito, T. A., J. T. Larsen, N. K. Smith, and J. T. Cacioppo. 1998. "Negative information weighs more heavily on the brain: The negativity bias in evaluative categorizations." *Journal of Personality and Social Psychology* 75: 887–900.

26. Atthowe, J. M. 1960. "Types of conflict and their resolution: A reinterpretation." *Journal of Experimental Psychology* 59: 1–9; Manne, S. L., K. L. Taylor, J. Dougherty, and N. Kemeny. 1997. "Supportive and negative responses in the partner relationship: Their association with psychological adjustment among individuals with cancer." *Journal of Behavioral Medicine* 20: 101–125.

27. Baumeister, R. F., and K. J. Cairns. 1992. "Repression and self-presentation: When audiences interfere with self-deceptive strategies." *Journal of Personality and Social Psychology* 62: 851–862.

28. David, J. P., P. J. Green, R. Martin, and J. Suls. 1997. "Differential roles of neuroticism, extraversion, and event desirability for mood in daily life: An integrative model of top-down and bottom-up influences." *Journal of Personality and Social Psychology* 73: 149–159.

29. Sheldon, K. M., R. Ryan, and H. T. Reis. 1996. "What makes for a good day?: Competence and autonomy in the day and in the person." *Personality and Social Psychology Bulletin* 22: 1270–1279.

30. Brickman, P., D. Coates, and R. Janoff-Bulman. 1978. "Lottery winners and accident victims: Is happiness relative?" *Journal of Personality and Social Psychology* 36: 917–927.

31. Cahill, C., S. P. Llewelyn, and C. Pearson. 1991. "Long-term effects of sexual abuse which occurred in childhood: A review." *British Journal of Clinical Psychology* 30: 117–130.

32. Ebbesen, E. B., G. L. Kjos, and V. J. Konecni. 1976. "Spatial ecology: Its effects on the choice of friends and enemies." *Journal of Experimental Social Psychology* 12: 505–518.

33. Czapinski, J. 1985. "Negativity Bias in Psychology: An Analysis of Polish Publications." *Polish Psychological Bulletin* 16: 27–44.

34. Ibid.

35. Riskey, D. R., and M. H. Birnbaum. 1974. "Compensatory effects in moral judgment: Two rights don't make up for a wrong." *Journal of Experimental Psychology* 103: 171–173.

36. Baumeister et al., 2001, 355.

37. Rozin, Paul, and Edward B. Royzman. 2001. "Negativity Bias, Negativity Dominance, and Contagion." *Personality and Social Psychology Review* 5(4): 296–320.

38. Hansen, C. H., and R. D. Hansen. 1988. "Finding the Face in the Crowd: An Anger Superiority Effect." *Journal of Personality and Social Psychology* 54: 917–924.

39. Schopenhauer, Arthur. 1844/1995. *The World as Will and Representation*, vol. 2. Translated by E.F.S. Payne. New York: Dover.

40. Rozin, P., L. Berman, and E. Royzman. 2001. *Posivity and Negativity Bias in Language: Evidence From 17 Languages*. Unpublished manuscript.

41. Frijda, N. H. 1986. *The Emotions*. Cambridge: Cambridge University Press.

42. The principle was made prominent in: Diamond, Jared. 1996. *Guns, Germs, and Steel*. New York: W. W. Norton.

43. Thompson, R. A. 1987. "Empathy and Emotional Understanding: The Early Development of Empathy." In N. Eisenberg and J. Strayer, eds. *Empathy and its Development*. New York: Cambridge University Press, 119–146.

44. Stevenson, H.N.C. 1954. "Status Evaluation in the Hindu Caste System." *Journal of the Royal Anthropological Institute of Great Britain and Ireland* 84: 45–65.

45. Miller, N. E. 1944. "Experimental studies of conflict." In J. McV. Hunt, ed. *Personality and the Behavior Disorders*, vol. 1. New York: Ronald Press, 435.

46. Rozin and Royzman, 2001, 306.

47. Pinker, Steven. 2017. "The Second Law of Thermodynamics." *Edge.org*, Annual Question: What Scientific Term or Concept Ought to Be More Widely Known? http://bit.ly/2hr7P2J

48. Pinker, Steven. 2018. *Enlightenment Now: The Case for Reason, Science, Humanism, and Progress*. New York: Penguin.

49. Shermer, Michael. 2011. *The Believing Brain*. New York: Henry Holt.

50. Diamond, Jared. 2012. *The World Until Yesterday: What Can We Learn from Traditional Societies?* New York: Viking Press, 243–275.

51. Eibach, Richard P., and Lisa K. Libby. 2009. "Ideology of the Good Old Days: Exaggerated Perceptions of Moral Decline and Conservative Politics." In *Social and Psychological Biases of Ideology and System Justification*, edited by J.T. Jost, A.C. Kay, and H. Thorisdottir, 402–423. New York: Oxford University Press.

52. Ladd, E. C. 1999. *The Ladd Report*. New York: Free Press.

53. LaFree, G. 1999. "Declining Violent Crime Rates in the 1990s: Predicting Crime Booms and Busts." *Annual Review of Sociology* 25: 145–168.

54. Sayer, L. C., S. M. Bianchi, and J. P. Robinson. 2004. "Are Parents Investing Less in Children? Trends in Mothers' and Fathers' Time with Children." *American Journal of Sociology* 110: 1–43.

55. National Campaign to Prevent Teen Pregnancy. 2003. *With One Voice 2003: America's Adults and Teens Sound Off about Teen Pregnancy*. Washington, DC.

56. Bork, Robert. 1996. *Slouching Towards Gomorrah: Modern Liberalism and American Decline*. New York: HarperCollins.

57. Murphy, A. R. 2005. "Augustine and the Rhetoric of Roman Decline." *History of Political Thought* 26: 586–606.

58. Hobbes, Thomas. 1651/1997. *Leviathan: Or the Matter, Forme, and Power of a Commonwealth Ecclesiasticall and Civil*, edited by Michael Oakeshott. New York: Simon and Schuster, 81.

59. Dupuy, Tina. 2016. "Once Upon a Time." *Skeptic* 21(2): 51–53.

CHAPTER 10

1. Cioran, E. M. 1960/1987. *History and Utopia*. Translated by Richard Howard. Chicago: University of Chicago Press, 81.

2. Eco, Umberto. 2013. *The Book of Legendary Lands*. New York: Rizzoli Ex Libris.

3. http://1.usa.gov/1SgZNFq

4. To name but a few across the ages: Plato's *The Republic* (360 B.C.E.), Cicero's *De Republica* (54 B.C.E.), Augustine's *The City of God* (C.E. 426), Al-Farabi's

The Virtuous City (c. 874–950), Thomas More's *Utopia* (1516), Tommaso Campanella's *The City of the Sun* (1623), Francis Bacon's *New Atlantis* (1627), James Harrington's *Commonwealth of Oceana* (1656), Gabriel de Foigny's *The Southern Land* (1676), Daniel Defoe's *Robinson Crusoe* (1719), Jonathan Swift's *Gulliver's Travels* (1726), Edward Bellamy's *Looking Backward* (1888), Theodor Hertzka's *Freeland* (1890), H. G. Wells's *The Time Machine* (1895), *A Modern Utopia* (1905), and *Men Like Gods* (1923), Charlotte Perkins Gilman's *Herland* (1915), B. F. Skinner's *Walden Two* (1948), Arthur C. Clarke's *Childhood's End* (1954), Ayn Rand's *Anthem* (1938) and *Atlas Shrugged* (1957), Ursula K. Le Guin's *The Lathe of Heaven* (1971), and Kim Stanley Robinson's The Mars trilogy (1992–96) and The Neanderthal Parallax trilogy (*Hominids, Humans, Hybrids* 2002–03).

5. See, for example: Davis, J. C. 1981. *Utopia and the Ideal Society: A Study of English Utopian Writing 1516–1700*. New York: Cambridge University Press.

6. De Camp, L. Sprague. 1970. *Lost Continent: The Atlantis Theme in History, Science, and Literature*. New York: Dover. See also: Ellis, Richard. 1998. *Imagining Atlantis*. New York: Alfred A. Knopf; Feder, Kenneth L. 2001. *Frauds, Myths, and Mysteries: Science and Pseudoscience in Archaeology*. New York: McGraw-Hill/Mayfield; Jordan, Paul. 2002. *The Atlantis Syndrome*. London: Sutton.

7. Albini, Andrea. 2012. *Atlantis: In the Textual Sea*. http://bit.ly/2gVXKKz

8. Utopia. Dystopia. *Oxford English Dictionary*. Oxford University Press. The genre sweeps across a broad range of things that can cause the collapse of a society: religion (Margaret Atwood's *The Handmaid's Tale*), politics (Ray Bradbury's *Fahrenheit 451*; George Orwell's *Animal Farm* and *1984*), economics (Ayn Rand's *Anthem* and *Atlas Shrugged*), ideology (Yevgeny Zamyatin's *We*, Franz Kafka's *The Trial*, Fritz Lang's *Metropolis*, Aldous Huxley's *Brave New World*, Arthur Koestler's *Darkness at Noon*, William Golding's *Lord of the Flies*), and especially science and technology (Philip K. Dick's *Minority Report* and *Do Androids Dream of Electric Sheep?*, Anthony Burgess's *A Clockwork Orange* and *Planet of the Apes*, Stanislaw Lem's *The Magellanic Cloud*, and Suzanne Collins's *The Hunger Games*).

9. Segal, Howard P. 2012. *Utopias: A Brief History from Ancient Writings to Virtual Communities*. Malden, MA: Wiley-Blackwell, 5.

10. Howard Segal makes this point in *Utopias*, op. cit., 5.

11. Gray, John. 2007. *Black Mass: Apocalyptic Religion and the Death of Utopia*. New York: Farrar, Straus and Giroux.

12. Emerson, Ralph Waldo. 1883. "Historic Notes of Life and Letters in New England." *The Works of Ralph Waldo Emerson: Lectures and Biographical Sketches*, vol. 10. Boston: Adamant Media Corporation, 327.

13. http://bit.ly/2nFQNgx

14. Quoted in: Tusman, Lee, ed. *Really Free Culture: Anarchist Communities, Radical Movements, and Public Practices*, 104. http://bit.ly/2mK3AiG

15. Clay, Alexa, 2017. "Utopia Inc." *Aeon*, February 28. http://bit.ly/2llj4Yo

16. For a history of Esalen see: Kripal, Jeffrey. 2007. *Esalen: America and the Religion of No Religion*. Chicago: University of Chicago Press.

17. Jones, Jim. 1978. *Transcript of Suicide Tape*. http://bit.ly/Ptom3S

18. Trotsky, Leon. 1924. "Literature and Revolution." http://bit.ly/2g2CqPi

19. Rossiianov, Kirill. 2002. "Beyond Species: Ilya Ivanov and His Experiments on Cross-Breeding Humans with Anthropoid Apes." In *Science in Context*. New York: Cambridge University Press, 277–316.

20. Walters, John J. 2013. "Communism Killed 94M in 20th Century, Feels Need to Kill Again." *Reason*, March 13. http://bit.ly/1nmmRkA

21. Chirot, Daniel, and Clark McCauley. 2006. *Why Not Kill Them All?: The Logic and Prevention of Political Murder*. Princeton, NJ: Princeton University Press, 143–144.

22. See, for example: Fitzgerald, Frances. 1981. *Cities on a Hill: A Journey Through Contemporary American Cultures*. New York: Simon and Schuster.

23. For a more positive assessment of some of the American utopian communities in the nineteenth century see: Holloway, Mark. *Utopian Communities in America, 1680–1880*. New York: Dover, 222–229.

24. The description of the failure of New Harmony was made by the individualist anarchist Josiah Warren in his 1856 *Periodical Letter II*. Quoted in: Brown, Susan Love, ed. 2002. *Intentional Community: Anthropological Perspective*. Albany, NY: State University of New York Press, 156.

25. Goldwater, Barry. 1964. *Acceptance Speech at the 28th Republican National Convention*. Arizona Historical Foundation. http://wapo.st/29RiHiq

26. Orwell, George. 1940. Review of *Mein Kampf* (unabridged translation). *New English Weekly*, March 21. In: Orwell, Sonia, and Ian Angus, eds. 1968. *Orwell: My Country Right or Left, 1940–1943*. Boston: Nonpareil Books, 12–14.

27. Foot, Phillipa. 1967. "The Problem of Abortion and the Doctrine of Double Effect." *Oxford Review* 5: 5–15. The extensive research utilizing the trolley car scenario has been summarized in many works, most recently in: Edmonds, David. 2013. *Would You Kill the Fat Man?* Princeton, NJ: Princeton University Press.

28. Paine, Thomas. 1795. *Dissertation on First Principles of Government*. Available online at: http://bit.ly/2qK9Tlr

29. Bury, J. B. 1920/1932. *The Idea of Progress: An Inquiry into Its Growth and Origin*. New York: Dover, 1–2.

30. Nisbet, Robert. 1980. *History of the Idea of Progress*. New York: Basic Books, 8–9.

31. Kelly, Kevin. 2011. "Protopia." KK.org, May 19. http://bit.ly/2h0jdSC

32. Shermer, Michael. 2015. *The Moral Arc*. New York: Henry Holt, 399.

33. Vollers, M. 2006. *Lone Wolf. Eric Rudolph: Murder, Myth and the Pursuit of an American Outlaw*. New York: HarperCollins, 302.

34. Isikoff, M. 2003. "Flushed From the Woods." *Newsweek*, June 9, 35.
35. Full text of Eric Rudolph's written statement. http://bit.ly/2fDFS6y
36. Kaczynski, Theodore J. 1995. *Industrial Society and Its Future*. *New York Times* and *Washington Post*, September 19.
37. Ibid. See also: Kaczynski, Theodore J. 2009. *The Road to Revolution*. Switzerland: Xenia.
38. Kennedy, Paul. 1987. *The Rise and Fall of the Great Powers*. New York: Random House, xvi.
39. I wrote about these in much greater depth for a chapter on "The New Revisionism" for the second edition of my coauthored book (with Alex Grobman) *Denying History: Who Says the Holocaust Never Happened and Why Do They Say It?* Shermer, Michael, and Alex Grobman. 2009. "The New Revisionism: Race, Politics, and the Unnecessary Good War." In *Denying History*, 2nd ed. Berkeley: University of California Press, 257–269.
40. Herman, Arthur. 1997. *The Idea of Decline in Western History*. New York: Free Press.
41. Kiernan, Ben. 2009. *Blood and Soil: A World History of Genocide and Extermination from Sparta to Darfur*. New Haven: Yale University Press. See also: Koonz, Claudia. 2005. *The Nazi Conscience*. Cambridge, MA: Harvard University Press.
42. Marx, Karl, and Friedrich Engels. 1851. *Manifesto of the Communist Party*. Available online: http://bit.ly/1DLXo9b
43. Gobineau, Arthur. 1853/1915. *The Essay on the Inequality of the Human Races*. London: William Heinemann. Full and searchable text available online: http://bit.ly/2fwVLrq
44. Stocking, George W. 1987. *Victorian Anthropology*. New York: Free Press, 67.
45. Gobineau, 1853, 206–211.
46. Field, Geoffrey. 1981. *The Evangelist of Race: The Germanic Vision of Houston Stewart Chamberlain*. New York: Columbia University Press, 421.
47. Stackelberg, R., and S. A. Winkle. 2002. *The Nazi Germany Sourcebook: An Anthology of Texts*. London: Routledge, 84–85.
48. Pletsch, Carl. 1992. *Young Nietzsche: Becoming a Genius*. New York: Free Press, 97.
49. Nietzsche, Friedrich. 1895/1968. *The Anti-Christ*. Translated by R. J. Hollingdale. New York: Penguin Books, 127.
50. Nietzsche, Friedrich. 1901/1968. *Will to Power*. Translated by Walter Kauffman. New York: Random House, 30.
51. Nietzsche, Friedrich. 1887/1969. *On the Genealogy of Morals*. Translated by Walter Kauffman. New York: Random House, 44.
52. Nietzsche, *The Anti-Christ*, 186.
53. Spengler, Oswald. *Decline of the West*. Translated by C. F. Atkinson. vol. 1. New York: Alfred A. Knopf, 21.

54. Herman, op. cit., 252.

55. Hitler, Adolf. 1925/1962. *Mein Kampf.* Translated by R. Manheim. New York: Houghton Mifflin, 289–290.

56. Quoted in: Hafner, Josh. 2016. "For the Record: For Trump, Everything's Going to be Alt-Right." *USA Today.* August 26. http://usat.ly/2bU000L Bannon made this declaration to: Posner, Sarah. 2016. "How Donald Trump's New Campaign Chief Created an Online Haven for White Nationalists." *Mother Jones.* August 22. http://bit.ly/2bH0DK0

57. Quoted in: Lombroso, Daniel, and Yoni Appelbaum. 2016. "'Hail Trump!' White Nationalists Salute the President-Elect." *Atlantic.* November 21. http://theatln.tc/2gbDPXY

58. Promotional video for the National Policy Institute: "Who Are We?" NPIAmerica.org/WhoAreWe

59. Quoted in: Cogan, Marin. 2016. "The Alt-Right Gives a Press Conference." *New York Magazine*, September 11. http://nym.ag/2cRj4QY

60. Michael, George. 2017. "The Rise of the Alt-Right and the Politics of Polarization in America." *eSkeptic*, February 1. http://bit.ly/2kUVyRR

61. Yuhas, Alan. 2015. "'Cuckservative': The Internet's Latest Republican Insult Hits Where It Hurts." *Guardian*, August 13. http://bit.ly/2l35HzB

62. Undonne, John. 2017. "What is the Alt Right?" AlternativeRight.com, February 1.

63. Ibid.

64. Ibid.

65. Barron, Christopher. 2016. "Donald Trump will be a friend, an ally and an advocate for the LGBT community." *Fox News.* November 15. http://fxn.ws/2fvOwDH

66. For a book-length treatment of Holocaust denial and this form of historical revisionism, particularly my attendance at this and many other such conferences, see: Shermer, Michael, and Alex Grobman. 2009. *Denying History: Who Says the Holocaust Never Happened and Why Do They Say It?,* 2nd ed. Berkeley: University of California Press.

67. Evans, Richard. 2001. *Lying about Hitler: History, Holocaust, and the David Irving Trial.* New York: Basic Books.

68. Cadwalladr, Carole. 2017. "Antisemite, Holocaust denier . . . yet David Irving Claims Fresh Support." *Guardian.* January 14. http://bit.ly/2jxjM55

69. Special Report. 2000. "Irving Taught his nine-month-old daughter racist ditty, libel trial told." *Guardian.* February 2. http://bit.ly/2kvHBKy

70. For Hitchens's full commentary on David Irving, and references to the above quote, see his essay "The Strange Case of David Irving," in his book *Love, Poverty, and War,* published by Nation Books in 2004.

71. Michael, George. 2009. "David Lane and the Fourteen Words." *Totalitarian Movements and Political Religions* 10(1): 41–59.

72. Downs, Caleb. 2016. "For white nationalists, Trump win a dream come true, says alt-right leader from Dallas." *Dallas News*, November 16. http://bit.ly/2fJ34xE

73. Mead, Walter Russell. 2017. "The Jacksonian Revolt." *Foreign Affairs*, January 20. http://fam.ag/2jhYnfB

74. Sohrab, Ahmari. 2016. *The New Philistines: How Identity Politics Disfigure the Arts*. London: Biteback Publishing.

CHAPTER 11

1. Samuelson, W., and R. J. Zeckhauser. 1988. "Status Quo Bias in Decision Making." *Journal of Risk and Uncertainty*, 1, 7–59.

2. 2013. "Living to 120 and Beyond: Americans' Views on Aging, Medical Advances and Radical Life Extension." Pew Research Center. August 6. http://pewrsr.ch/1ZsCUPR

3. Hitchens, Christopher. 2010. "Topic of Cancer." *Vanity Fair*, August. http://bit.ly/1UaIZA3

4. Tversky, Amos, and Daniel Kahneman. 1973. "Availability: A Heuristic for Judging Frequency and Probability." *Cognitive Psychology* 5: 207–232.

5. Glassner, Barry. 1999. *The Culture of Fear: Why Americans Are Afraid of the Wrong Things*. New York: Basic Books.

6. http://bit.ly/1c9a3vO

7. For U.S. death data see: National Vital Statistics Reports. 2016. Deaths: Final Data for 2013. CDC, vol. 64, no. 2, 10. http://1.usa.gov/1GEJ0TN

8. J. Nielsen et al. 2016. "Eye lens radiocarbon reveals centuries of longevity in the Greenland shark (*Somniosus microcephalus*)." *Science* 353:702–704.

9. Tosato, Matteo, Valentina Zamboni, Alessandro Ferrini, and Matteo Cesari. 2007. "The Aging Process and Potential Interventions to Extend Life Expectancy." *Clinical Interventions in Aging* 2(3): 401–412. http://bit.ly/21uJquc

10. Hayflick, Leonard. 2000. "The Future of Aging." *Nature* 408: 267–269.

11. Medawar, Peter. 1952. *An Unsolved Problem of Biology*. Published for the College by H. K. Lewis. http://bit.ly/2auTYTu

12. Hayflick, Leonard. 2007. "Biological Aging is No Longer an Unsolved Problem." *Annals of the New York Academy of Sciences*, April. http://bit.ly/1MaUe81

13. McCall, M. R., and B. Frei. 1999. "Can Antioxidant Vitamins Materially Reduce Oxidative Damage in Humans?" *Free Radical Biological Medicine* 26: 1034–1053.

14. Trut, L. N. 1995. "Domestication of the Fox: Roots and Effects." *Scientifur* 19: 11–18.

15. Williams, G. C. 1957. "Pleiotropy, Natural Selection, and the Evolution of Senescence." *Evolution* 11: 398–411. See also: Eaton, S. G., et al. 1994. "Women's Reproductive Cancers in Evolutionary Context." *Quarterly Review of Biology* 69: 353–367.

16. Hayflick, Leonard. 1965. "The Limited in Vitro Lifetime of Human Diploid Cell Strains." *Experimental Cell Research* 37: 614–636.

17. Fossel, M. 1998. "Telomerase and the Aging Cell: Implications for Human Health." *JAMA* 279: 1732–1735.

18. Shay, J. W., and W. E. Wright. 1996. "Telomerase Activity in Human Cancer." *Current Opinion in Oncology* 8: 66–71.

19. Funk, W. D., C. K. Wang, et al. 2000. "Telomerase Expression Restores Dermal Integrity to In Vitro–Aged Fibroblasts in a Reconstituted Skin Model." *Experimental Cell Research* 258: 270–278.

20. Bryan, T. M., A. Englezou, et al. 1995. "Telomere Elongation in Immortal Human Cells without Detectable Telomerase Activity." *EMBO J.* 14: 4240–4248.

21. Ornish, Dean, Jue Lin, June M. Chan, et al. 2013. "Effect of Comprehensive Lifestyle Changes on Telomerase Activity and Telomere Length in Men with Biopsy-Proven Low-Risk Prostate Cancer: 5-Year Follow-Up of a Descriptive Pilot Study." *The Lancet Oncology* 14/11: 1112–1120.

22. Blackburn, Elizabeth, and Elissa Epel. 2017. *The Telomere Effect*. New York: Grand Central Publishing.

23. De Grey, Aubrey, and Michael Rae. 2008. *Ending Aging: The Rejuvenation Breakthroughs that Could Reverse Human Aging in Our Lifetime*. New York: St. Martin's Press. See also: de Grey, Aubrey. *The Mitochondrial Free Radical Theory of Aging*. New York: Cambridge University Press.

24. Stated in the television show: *Aux Frontieres de I'mmortalite*. 2008. Gerald Calliat, director. November 16.

25. Best, Ben. 2013. "Interview with Aubrey de Grey, Ph.D." *Life Extension Magazine*.

26. The SENS Research Foundation website has a good summary of the seven aging issues here: http://bit.ly/1bt48uG

27. 2012. "Is Defeating Aging Only a Dream?" *Technology Review*.

28. Warner, H., J. Anderson, S. Austad, et al. 2005. "Science Fact and the SENS Agenda: What Can We Reasonably Expect from Aging Research?" *EMBO Reports*, November, 1006–1008.

29. Quoted in: SENS Research Foundation FAQ. SENS Research Foundation. Interestingly, this line was superseded with this: "no currently available medical intervention or lifestyle choice has been shown to affect the basic human aging process." http://bit.ly/29XWBj4

30. Gifford, Bill. 2016. "Is 100 the New 80?" *Scientific American*, December. http://bit.ly/2gUANqB

31. Weintraub, Karen. 2016. "Aging Is Reversible—at Least in Human Cells and Live Mice." *Scientific American*, December 15.

32. Olshansky, S. Jay, Leonard Hayflick, and Bruce A. Carnes. 2002. "No Truth to the Fountain of Youth." *Scientific American*, June.

33. De Gaetano, G., S. Costanzo, et al. 2016. "Effects of Moderate Beer Consumption on Health and Disease: A Consensus Document." *Nutrition Metabolism Cardiovascular Disease*, June, 443–467. The authors caution that their results do not include binge beer drinking, which lead to "deleterious effects on the human body, with increased disease risks on many organs and is associated to significant social problems such as addiction, accidents, violence and crime."

34. Nuland, Sherwin B. 1993. *How We Die: Reflections on Life's Final Chapter*. New York: Random House, 267.

35. Kirkwood, Thomas B. 1977. "Evolution of Aging." *Nature* 270: 301–304.

36. Austad, S. N. 1993. "Retarded Senescence in an Insular Population of Virginia Opossums (*Dipelphis virginiana*)." *Journal of Zoology* 229: 695–708.

37. Bribiescas, Richard G. *How Men Age: What Evolution Reveals about Male Health and Mortality*. Princeton, NJ: Princeton University Press, 25.

38. Goldsmith, T. C. 2008. "Aging, Evolvability, and the Individual Benefit Requirement: Medical Implications of Aging Theory Controversies." *Journal of Theoretical Biology* 252: 764–768.

39. Fabian, Daniel, and Thomas Flatt. 2011. "The Evolution of Aging." *Nature Education Knowledge* 3(10): 9. http://go.nature.com/1S3DAVZ

40. Dawkins, Richard. 1976. *The Selfish Gene*. New York: Oxford University Press.

41. Dawkins says he considered titling his famous book *The Immortal Gene* instead of *The Selfish Gene*.

42. http://bit.ly/2ddwzaQ

43. Smolin, Lee. 1997. *The Life of the Cosmos*. New York: Oxford University Press; Liddle, Andrew, and Jon Loveday. 2009. *The Oxford Companion to Cosmology*. New York: Oxford University Press; Weinberg, Stephen. 2008. *Cosmology*. New York: Oxford University Press.

44. Dyson, Freeman. 1979. "Time Without End: Physics and Biology in an Open Universe." *Reviews of Modern Physics* 51(3), July. http://bit.ly/2rqneTo

45. Pollack, James, and Carl Sagan. 1993. "Planetary Engineering." In *Resources of Near Earth Space*. Lewis, J., M. Matthews, and M. Guerreri (eds.). Tucson: University of Arizona Press; Niven, Larry. 1990. *Ringworld*. New York: Ballantine; Stapledon, Olaf. 1968. *The Starmaker*. New York: Dover.

46. I first proposed this idea in: Shermer, Michael. 2002. "Shermer's Last Law." *Scientific American*, January, 33. I developed it further in my book *The Believing Brain*, op. cit. It is an acknowledged derivation from Arthur C. Clarke's Third Law: "Any sufficiently advanced technology is indistinguishable from magic." My gambit is based on three observations and four deductions, the fourth involving far future humans, added here for the first time:

Observation I. Biological evolution is glacially slow compared to cultural and technological evolution.

Observation II. The cosmos is very big and space is very empty, so the probability of making contact with an ETI is remote.

Deduction I. The probability of making contact with an ETI who is only slightly more or less advanced than us is virtually nil. Any ETIs we would encounter will either be way behind us or way ahead of us.

Observation III. Science and technology have changed our world more in the past century than it changed in the previous hundred centuries. Moore's Law of computer power doubling every twelve months applies to dozens of other technologies. If continued at this pace the world will change more in the next century than it has in the previous thousand centuries.

Deduction II. Extrapolate these trend lines out tens of thousands, hundreds of thousands, or even millions of years—mere eye blinks on an evolutionary time scale—and we arrive at a realistic estimate of how far advanced an ETI will be.

Deduction III. If today we can engineer genes, clone mammals, and manipulate stem cells with science and technologies developed in only the last fifty years, think of what an ETI could do with fifty thousand years of equivalent powers of progress in science and technology. For an ETI who is a million years more advanced than we are, engineering the creation of planets and stars may be entirely possible. And if universes are created out of collapsing black holes—which some cosmologists think is possible—it is not inconceivable that a sufficiently advanced ETI could even create a universe by triggering the collapse of stars into a black hole.

Deduction IV. All of these deductions apply to far future humans, particularly once the technological singularity is reached.

47. Mechelin, Leopold Henrik Stanislaus. 1894. *Finland in the Nineteenth Century.* Helsingfors: F. Tilgmann, 274.
48. Mayr, Ernst. 1957. "Species Concepts and Definitions," in *The Species Problem.* Washington D.C.: Amer. Assoc. Adv. Sci. Publ., no. 50. See also: Mayr, Ernst. 1988. *Toward a New Philosophy of Biology.* Cambridge, MA: Harvard University Press.
49. Through boldness to the stars.

CHAPTER 12

1. Rushdie, Salman. 1999. "Imagine No Heaven," *Guardian*, October 15. http://bit.ly/1SKoeGF
2. Crane, Stephen. 1899. No Title. In *War Is Kind and Other Lines.* http://bit.ly/28CXuTC

3. Diana Nyad on Oprah Winfrey's *Super Soul Sunday*, October 6, 2013: http://bit.ly/1UiKUof

4. Valdesolo, Piercarlo, and Jesse Graham. 2013. "Awe, Uncertainty, and Agency Detection." *Psychological Science*, November 18. http://bit.ly/1YBosGK

5. Personal correspondence, December 13, 2013.

6. See the full explanation of patternicity and agenticity in chapters 4 and 5 of: Shermer, Michael. 2011. *The Believing Brain*. New York: Henry Holt.

7. Personal correspondence, December 7, 2013.

8. Nyad, Diana. 2015. *Find a Way: One Wild and Precious Life*. New York: Alfred A. Knopf, 239.

9. Modern astronomers and science writers such as Lawrence Krauss and Neil deGrasse Tyson are fond of saying that we're made of "star stuff" or "star dust," which most people think is being channeled from one of the greatest scientists and science writers of all, Carl Sagan, but Carl's widow, Ann Druyan, informs me that the original source of the phrase was Harlow Shapley. Phone interview, June 11, 2014.

10. Tooby, John, Leda Cosmides, and H. Clark Harrett. 2003. "The Second Law of Thermodynamics Is the First Law of Psychology." *Psychological Bulletin* 129(6): 858–865.

11. Khayyám, Omar. 1120/2009. *The Rubaiyat*, 5th ed. Translated by Edward FitzGerald. Oxford: Oxford University Press, 30.

12. Shakespeare, William. 1603. *Hamlet*. Act 1, scene 3. http://bit.ly/2661T2P

13. Baumeister, Roy F., Kathleen D. Vohs, Jennifer Aaker, and Emily N. Garbinsky. 2013. "Some Key Differences between a Happy Life and a Meaningful Life." *Journal of Positive Psychology* 8(6): 505–516.

14. Baumeister, Roy. 2013. "The Meanings of Life." *Aeon*. http://bit.ly/2lnSuzv

15. Frankl, Viktor E. 1946. *Man's Search for Meaning*. Boston: Beacon, 37–38.

16. Vail, Kenneth E., and Jacob Juhl. 2015. "An Appreciative View of the Brighter Side of Terror Management Processes." *Social Sciences* 4: 1020–1045. http://bit.ly/2lwe7A2

17. Ibid.

18. See also: Vail, Kenneth E., Jacob Juhl, Jamie Arndt, Matthew Vess, Clay Routledge, and Bastiaan T. Rutjens. 2012. "When Death Is Good for Life: Considering the Positive Trajectories of Terror Management." *Personality and Social Psychology Review* 16(4). http://bit.ly/2lRhZJ1

ACKNOWLEDGMENTS

To Pat Linse for rendering some of the illustrations for this book, and for her friendship and professional partnership in skepticism for over a quarter century.

To my general editor, Serena Jones, editorial assistant Madeline Jones, production editors Molly Bloom and Olivia Croom, and copy editor Emily DeHuff, with gratitude for turning my words into a polished and coherent book. And to the rest of the wonderful team at Henry Holt/Macmillan for launching this book into the world: Paul Golob, Maggie Richards, Carolyn O'Keefe, and Jessica Wiener.

To my agents, Katinka Matson, John Brockman, Max Brockman, Russell Weinberger, and the staff of Brockman, Inc., which not only manages the largest stable of science authors in the world but also nourishes the most stimulating online salon of scientists, philosophers, and scholars for the purpose of sharing their thoughts and generating new ideas about the world's most interesting topics through the Edge.org web community.

To John Rael and Randy Olson for their brilliant video production of our Science Salon series and additional film shorts that deliver science and skepticism in a powerful visual medium.

To Alexander Pietrus-Rajman for expanding our horizons throughout Europe as well as new social media, and for bringing fresh new ideas into an organization now a quarter century old.

To Jayde Lovell, Rebecca Gill, and the team at the Reagency Public

Relations firm that helped launch the Skeptics Society and *Skeptic* magazine to new heights and thereby promote science and skepticism, the dual engines that drive the modern world.

Much of this book was written at my office, so I also wish to recognize the many fine people working at or associated with the Skeptics Society and *Skeptic* magazine, including Nicole McCullough, Priscilla Loquellano, Daniel Loxton, William Bull, Jerry Friedman, and most especially my partner, Pat Linse. Our many volunteers that make such an organization run smoothly deserve acknowledgment: senior editor Frank Miele; senior scientists David Naiditch, Bernard Leikind, Liam McDaid, Claudio Maccone, and Thomas McDonough; contributing editors Tim Callahan, Harriet Hall, and Carol Tavris; editor Sara Meric; photographer David Patton and videographer Brad Davies; and our many volunteers: Cliff Caplan, Michael Gilmore, and Diane Knudtson.

To David and Jackie Naiditch for opening their home to our Science Salon and thereby enriching so many people with such a beautiful setting for discussing science and skepticism.

To my lecture agents, Scott Wolfman and Diane Thompson at Wolfman Productions, for their contribution in bringing science and skepticism to the speaker's circuit.

To my colleagues (and friends) Anondah Saide and Kevin McCaffree, who made important contributions to the analysis of death-row inmates' final statements in chapter one.

To Danielle Struppa, president of Chapman University, for nourishing an environment of open dialogue and for protecting all forms of free speech at this wonderful campus, and for enabling me to teach first-year students how to think critically through Skepticism 101.

To the editorial and art departments at *Scientific American,* the longest continuously published magazine in American history, who have allowed me to contribute every month since April 2001: Mariette DiChristina, Fred Guterl, Michael Lemonick, Christi Keller, Aaron Shattuck, Michael Mrak, and especially Izhar Cohen for his remarkable illustrations that turn my words into striking visual images.

To my daughter, Devin, with love always.

To my wife, Jennifer, for everything and forever, and to our son, Vincent Richard Walter Shermer, to whom this book is dedicated.

INDEX

Page numbers in *italics* refer to illustrations.

ABOUT THE AUTHOR

DR. MICHAEL SHERMER is the founding publisher of *Skeptic* magazine, a monthly columnist for *Scientific American*, and Presidential Fellow at Chapman University. He is the author of *The Moral Arc*, *The Believing Brain*, *Why People Believe Weird Things*, *Why Darwin Matters*, *The Mind of the Market*, *How We Believe*, and *The Science of Good and Evil*. He has been a college professor since 1979, having also taught at Occidental College, Glendale College, and Claremont Graduate University. As a public intellectual he regularly contributes Opinion Editorials, book reviews, and essays to *The Wall Street Journal*, *The Los Angeles Times*, *Science*, *Nature*, and other publications. Dr. Shermer received his B.A. in psychology from Pepperdine University, M.A. in experimental psychology from California State University, Fullerton, and his Ph.D. in the history of science from Claremont Graduate University. He appeared on such shows as *The Colbert Report*, *20/20*, *Dateline*, *Charlie Rose*, and *Larry King Live* (but, proudly, never *Jerry Springer*!). His 2 TED talks, seen by many millions, were voted in the top 100 of the more than 2000 TED talks, and he was one of the select few to deliver a TED All Star talk.